Contents

Introduction

Purpose

This book is a resource for providing additional challenging mathematics to that already included in the Abacus 6 materials, to further extend more-able pupils.

It is the responsibility of all teachers to ensure that children are given every opportunity to realise their full mathematical potential. It is important, therefore, that the development of the more-able child is not restricted. There are numerous opportunities within Abacus 6 for teachers to select tasks appropriate for the more-able child. The Challenge Book provides additional suggestions for such opportunities with a range of structured, challenging and purposeful activities.

Structure

The book is divided into sections relating to Units or groups of Units within Abacus 6.

It is assumed that most children for whom this material is appropriate will have satisfactorily completed the related Abacus Unit(s), and would benefit from further related activity in order to extend their mathematical development.

The suggested Challenges provide the teacher with a wide selection of activities within each Unit for the more-able children. It is not intended that they are all used, but that the teacher will select from this bank of ideas those activities which are appropriate.

Using the Challenges

The activities for each Unit or group of Units are contained in one double-page spread. The left-hand page provides Teacher's Notes relevant to the particular Unit, and the right-hand page is always a photocopiable 'Challenge Master'. The activities are easy to administer, require minimum use of materials, and minimum teacher input.

The Teachers' Notes are arranged under the following headings:

Skills summary

A summary of the key mathematical skills developed within and beyond the particular Abacus 6 Unit.

Core links

Notes with details of how to further extend activities within the existing Abacus 6 Textbook and Photocopy Masters.

Introduction

Challenge master

Notes relating to the relevant Challenge Master, with details of how to further extend this. The photocopiable Challenge Masters themselves are presented directly to the child and always appear opposite the relevant page of Teacher's Notes.

Challenge activities

Practical activities to further extend the child's understanding of a particular mathematical topic, often using simple materials, e.g. number cards (many of which can be found in the Resource Bank).

Roughly half of all the activities are designed to develop the children's mathematical 'breadth', in which the children seek to broaden their understanding of a piece of mathematics but at the **same** level.

The remainder of the activities are designed to develop the children's mathematical 'depth', and are clearly labelled ➊. In these activities children extend their study of a particular piece of mathematics, but at a **higher** level.

The suggested activities are often open-ended and of a problem-solving nature, geared towards the development of the children's process skills. It is anticipated that this Challenge Book will develop the child's ability to 'use and apply mathematics' at a higher level.

For many of the Challenge Masters, children can record their work on the sheet. Some Challenge Master activities require the children to record their work on separate sheets or in an exercise book. These sections are highlighted by a recording 'icon':

Children should be encouraged to record their work clearly and systematically.

Place-value and rounding

Skills summary

- To recognise place-value in numbers with more than four digits
- To read numbers with more than four digits written in words, and write the numerals
- To write numbers with more than four digits in words, given the numerals
- To rehearse rounding a 4-digit number to its nearest 10, nearest 100 and nearest 1000
- To round a number with more than four digits to its nearest 10, 100, ... , million
- To recognise place-value in numbers with up to three decimal places
- To round a 1-place decimal number to its nearest whole number
- To round a 2-place decimal number to its nearest whole number, tenth
- To round a 3-place decimal number to its nearest whole number, tenth, hundredth
- To estimate the result of a calculation by approximating (calculating with rounded numbers)

Core links

Number Textbook 1 page 3, top

- Calculate the total number of fans from all the games, then calculate the percentage of each of the total.

Number Textbook 1 page 5, bottom

- Calculate the cost of each hat if they are reduced by 20%.

Number Textbook 2 page 29, top

- Estimate the area of each rectangle as accurately as you can, then calculate it using a calculator. Compare the two.

Number Textbook 2 page 30, top

- Calculate the average (mean) cost of season tickets, then how much each club is above or below the mean.

Photocopy Master 2

- Change the game to rounding to the nearest hundred, and scoring points to match the number of hundreds.

Challenge Master 1

Extension Activity

- Using the four cards, investigate how many different decimal numbers can be created, and then how many different nearest tenths.

Challenge Activity

Activity 1

- Use a set of number cards (0 to 9) and a counter for a decimal point. Shuffle the cards and lay out four to make a 3-place decimal number, e.g. 4·379. Say the number, then round it to its nearest hundredth, tenth and whole number. Repeat for different numbers.

Name ____________________________

Rounding

Use three of these four cards to make decimal numbers.

3  9 4  6

Make numbers that round to the given numbers to the nearest tenth.

1 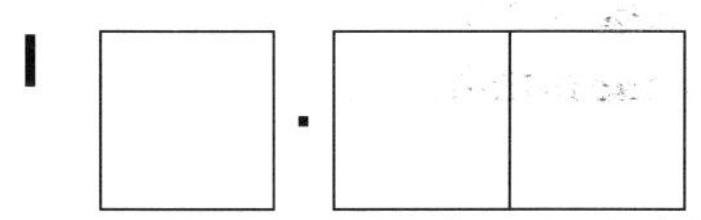 □ · □ □ → 4·6

2   □ · □ □ → 9·5

3 □ · □ □ → 3·6

4  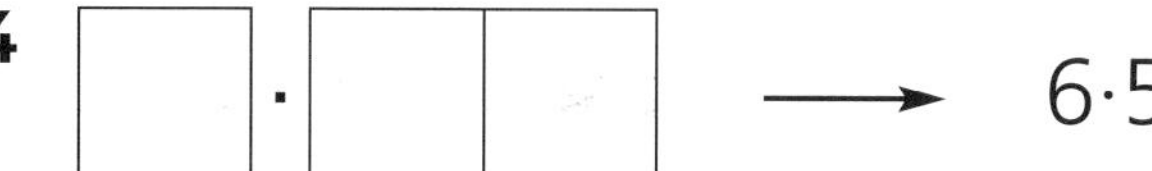 □ · □ □ → 6·5

5 □ · □ □ → 4·0

6 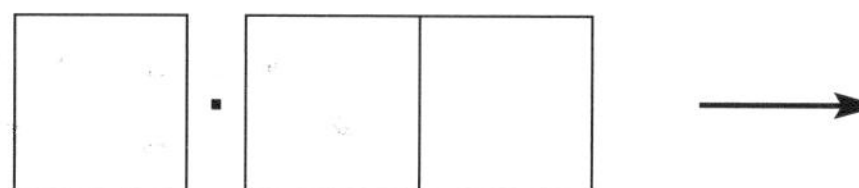 □ · □ □ → 3·7

7 □ · □ □ → 4·9

8 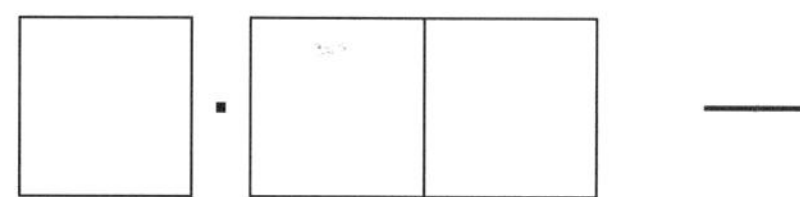 □ · □ □ → 6·9

9   □ · □ □ → 3·5

10 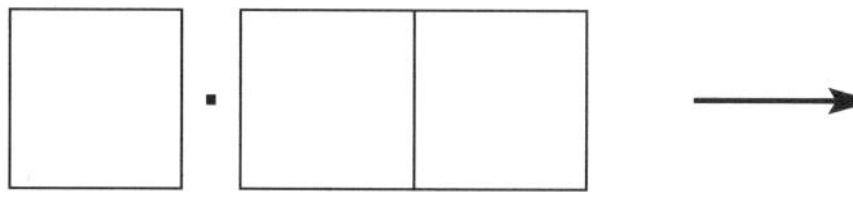 □ · □ □ → 4·7

11 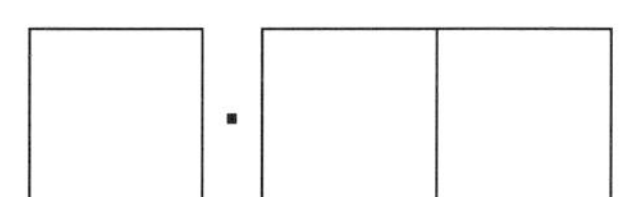 □ · □ □ → 6·3

12 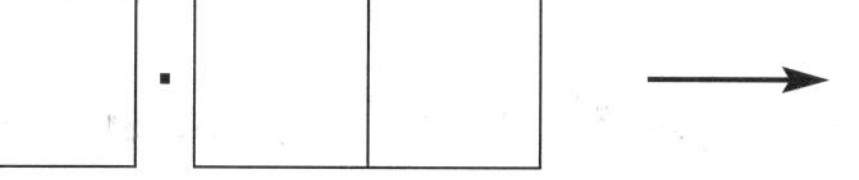 □ · □ □ → 4·4

13 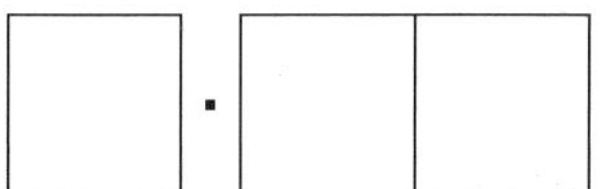 □ · □ □ → 9·4

14  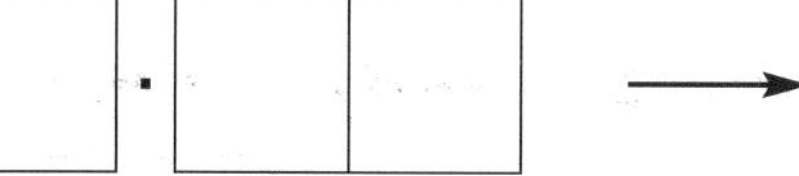 □ · □ □ → 9·3

15 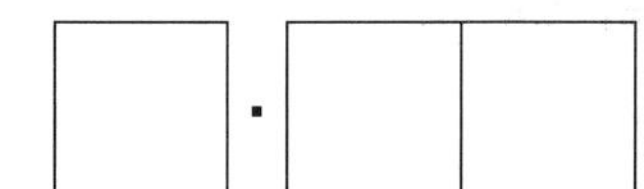 □ · □ □ → 3·9

16 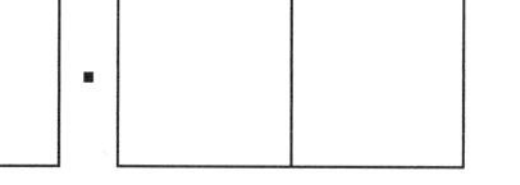 □ · □ □ → 5·0

 Choose your own four number cards.
Investigate how many different 'nearest tenths' you can make with them.

N2 N15 N16 Place-value

Multiplying and dividing by 10

Skills summary

- To multiply a whole number by 10, by 100, by 1000
- To divide a multiple of 10 (e.g. 280) by 10
- To divide a multiple of 100 (e.g. 5700) by 100
- To multiply numbers by multiples of 100
- To multiply a 1-place or 2-place decimal number by 10
- To multiply a 1-place or 2-place decimal number by 100
- To divide a decimal number by 10, by 100
- To multiply whole numbers and decimal numbers by any power of 10

Core links

Number Textbook 1 page 8, top

- Calculate the total capacity of all nine containers. Approximately how many pints will be needed to fill them all? (1 litre is approximately $1\frac{3}{4}$ pints.)

Number Textbook 1 page 49, bottom

- Find each total cost if a discount of 15% is given.

Number Textbook 1 page 51, top

- Write each distance to the nearest mile (50 miles is approximately 80 km).

Number Textbook 1 page 52, top

- Write, as accurately as you can, the proportional height of each animal compared with the dog.

Photocopy Master 4, bottom

- Calculate the cost of each item if they are reduced by 35%.

Photocopy Master 32

- Write each length to the nearest foot.

Challenge Master 2

Extension Activities

- Round all decimal answers to the nearest tenth, hundredth.
- Devise fast methods for multiplying by other numbers using halving and doubling, e.g. ×16, ×60.
- Extend to dividing by 20, for example, numbers whose last digit is odd, e.g. 6·41 ÷ 20. Half of 6·41 = 3·205. Dividing this by 10 gives 0·3205.

Name ______________________________

Fast multiplying and dividing

A fast method of multiplying by 400, for example, is to multiply by 4 (by doubling, then doubling again), then multiply by 100.

For example, 5·6 x 400
double 5·6 = 11·2
double 11·2 = 22·4
22·4 x 100 = 2240

Use doubling to help you with these multiplications:

1 $0{\cdot}81 \times 20 =$ ____________

2 $1{\cdot}75 \times 40 =$ ____________

3 $15{\cdot}6 \times 200 =$ ____________

4 $0{\cdot}136 \times 4000 =$ ____________

5 $4{\cdot}32 \times 80 =$ ____________

6 $2{\cdot}95 \times 400 =$ ____________

7 $0{\cdot}076 \times 2000 =$ ____________

8 $1{\cdot}09 \times 800 =$ ____________

A fast method of dividing by 80, for example, is to divide by 8 (by halving, halving again, then halving again), then divide by 10.

For example, 5·44 ÷ 80
half of 5·44 = 2·72
half of 2·72 = 1·36
half of 1·36 = 0·68
0·68 ÷ 10 = 0·068

Use halving to help you with these divisions:

9 $7{\cdot}82 \div 20 =$ ____________

10 $58{\cdot}6 \div 40 =$ ____________

11 $9320 \div 400 =$ ____________

12 $93{\cdot}2 \div 80 =$ ____________

13 $196{\cdot}4 \div 200 =$ ____________

14 $581{\cdot}8 \div 2000 =$ ____________

15 $636 \div 800 =$ ____________

16 $4656 \div 4000 =$ ____________

Multiplication/division

Skills summary

- To recall multiplication facts ×2, ×3, ×4, ×5, ×10, and related division facts
- To recall multiplication facts ×6, ×7, ×8, ×9, and related division facts
- To recall all multiplication facts up to 10 × 10, and related division facts
- To recall or derive quickly doubles of numbers from 1 to 100 and corresponding halves
- To recall or derive quickly doubles of multiples of 10 to 1000, and corresponding halves
- To recall or derive quickly doubles of multiples of 100 to 10000, and corresponding halves
- To recall or derive quickly doubles of 1-place decimal numbers, and corresponding halves
- To recall or derive quickly doubles of 2-place decimal numbers, and corresponding halves

Core links

Number Textbook 1 page 10, top

- How many boxes are needed of each item to pack at least 100 altogether?

Number Textbook 1 page 12, top

- Calculate, to the nearest whole number, the average (mean) number that each child receives.

Number Textbook 1 page 14, top

- Write the ratio of points for the teams in each question. Write the ratios in their simplest form.

Number Textbook 1 page 15, top

- Write the new prices for a different reduction, e.g. 15%, 35%.

Number Textbook 1 page 16, top

- Write each answer to the nearest mile (5 miles is approximately 8 kilometres).

Photocopy Master 9

- Introduce a second '×' card, and make sentences like these: 2 × 6 × 7 = 84.
- Use a '÷' card instead of a '×' card, and make division sentences.

Challenge Master 3

Extension Activity

- Investigate all the different possible ways of splitting a strip of ten in two, then splitting it into three, then four, and so on. Extend this to strips of different lengths.

Challenge Activity

- Use a set of number cards (0 to 9) and a counter for a decimal point. Shuffle the cards and lay out a pair to make a 1-place decimal number, e.g. 5·6. Say double the number, then half the number. Repeat and then extend to numbers with two decimal places.

Name ______________________________

N3 N4

Splitting the strip

Start with a strip of 10.

Split the strip into pieces and multiply the parts together.

3 × 1 × 6 = 18

2 × 2 × 3 × 3 = 36

6 × 4 = 24

2 × 5 × 3 = 30

Investigate the answer for different splits.
Which split gives the largest answer?
Try starting with a different strip length.

Multiplication/division

Skills summary

- To multiply by near multiples of 10, by multiplying by 10, then adjusting
- To multiply by near multiples of 100, by multiplying by 100, then adjusting
- To multiply by near multiples of 50, by multiplying by 100, halving, then adjusting
- To multiply by any near multiple of 10, by multiplying by the multiple of 10, then adjusting
- To multiply two numbers by first doubling one and halving the other
- To derive one multiplication table by doubling another, e.g. double ×8 for ×16
- To derive hard multiplication facts by doubling part of easier facts
- To multiply/divide by 5 by multiplying/dividing by 10, then halving/doubling
- To multiply/divide by 50 by multiplying/dividing by 100, then halving/doubling
- To multiply by 25 by multiplying by 100, then halving and halving again
- To recognise different methods of multiplying, and use these as a check

Core links

Number Textbook 1 page 17, top

- How much would each child need to save weekly for a year, for each trip?

Number Textbook 1 page 18, top

- Calculate what price the tickets for each concert will be, based on the larger seating arrangement, to take £100 000.

Number Textbook 1 page 19, top

- Write multiplications with answers close to 2800. How close is each answer?

Number Textbook 1 page 20, bottom

- Calculate how many tickets for each event you can buy for £30.

Number Textbook 1 page 21, bottom

- How many of each item can you buy for a given amount, e.g. £15?

Number Textbook 1 page 22, top

- Calculate how many boxes can be packed in a day, i.e. 24 hours.

Photocopy Master 11

- Invent your own multiplication tables, including 3-digit numbers which are not multiples of 100, e.g. 340, 271.

Photocopy Master 13

- Extend the activity to 3-digit numbers, rolling the dice three times.

Challenge Master 4

Extension Activity

- Write a description of how the Egyptian method of multiplication works.
- Try the Egyptian method using decimal numbers.

Challenge Activity

- Using a set of number cards (0 to 9), deal out two 2-digit numbers. Write as many different ways as you can find of performing the multiplication.

Name ______________________________ **N5** **N6**

Egyptian multiplication

This method of multiplying was used by the ancient Egyptians.
Start with a multiplication of two 2-digit numbers.

- Write each number at the top of a column.
- Take the number on the left and halve it. Round the answer down to a whole number and write it underneath.
- Keep halving, rounding down each time, until you reach 1.
- Now double the number on the right as many times.
- Cross out any numbers in the right-hand column that are opposite an even number.
- Add the remaining numbers on the right for the answer.

19 × 27

19	27
9	54
4	108
2	216
1	432

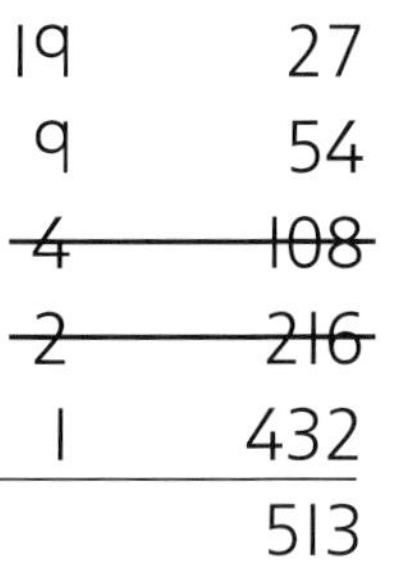

19 x 27 = 513

Try the Egyptian method on these multiplications.

1 17 × 45 **2** 26 × 33 **3** 18 × 51

4 35 × 21 **5** 46 × 53 **6** 62 × 34

Check the results using a different method.

Fractions

Skills summary

- To find a fraction (unit numerator) of a quantity, e.g. $\frac{1}{6}$ of 20
- To find a fraction (non-unit numerator) of a quantity, by first finding the unit numerator fractions, e.g. $\frac{3}{5}$ of 40 by finding $\frac{1}{5}$
- To recognise what fraction, in its lowest terms, one amount is of another
- To find a percentage (multiple of 10%) of an amount
- To find any percentage of an amount
- To divide a quantity into a given ratio
- To increase and decrease an amount by a given percentage

Core links

Number Textbook 1 page 23, top

- Write each number as a percentage, to the nearest whole number.

Number Textbook 1 page 24, top

- Calculate how much money the village fête makes in total, after giving some to charity.

Number Textbook 1 page 25, top

- Write each fraction as a percentage, to the nearest whole number.

Challenge Master 5

Extension Activity

- Devise your own fraction wheels. Choose an amount for the centre, then choose four fractions of this amount, and calculate the outside numbers of the wheel.

Challenge Activity

- Use a set of number cards (0 to 9) to deal out two 2-digit numbers. Calculate what fraction, in its lowest terms, the smaller amount is of the larger. Repeat for different pairs of numbers, recording each pair and its fraction.

Name ______________________

N7

Fraction wheels

Write what fraction each amount on the outside of the wheel is of the amount in the centre. Write each fraction in its lowest terms.

For example, £25 out of £45 is $\frac{25}{45} = \frac{5}{9}$.

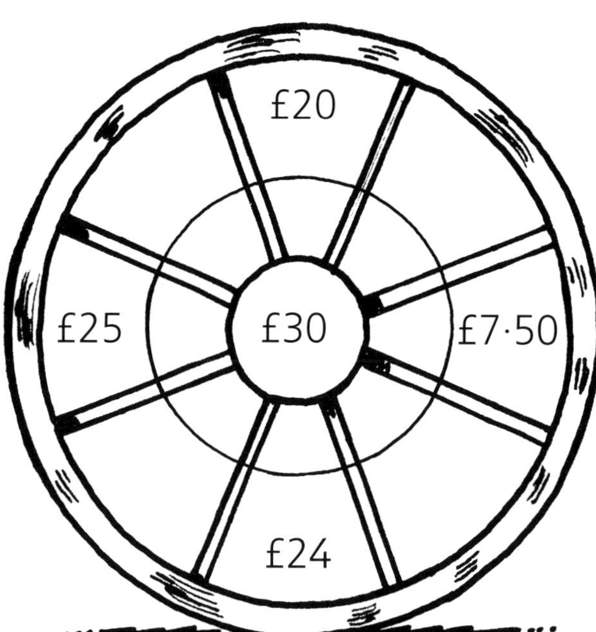

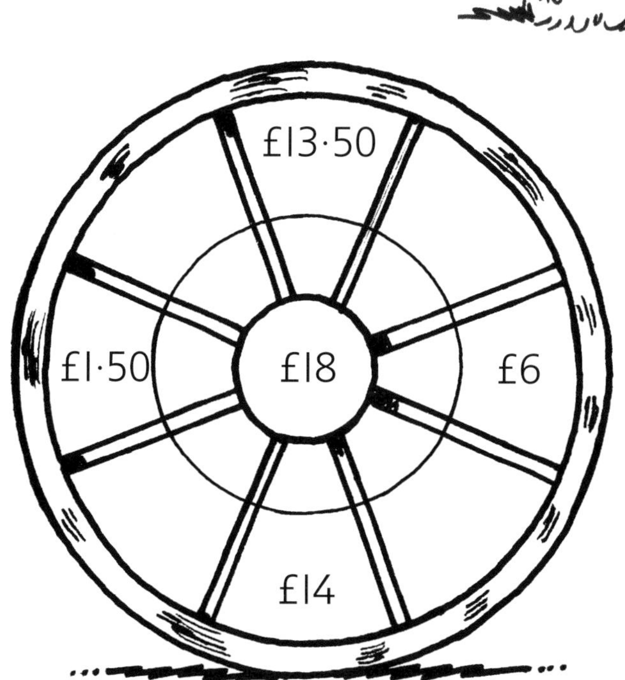

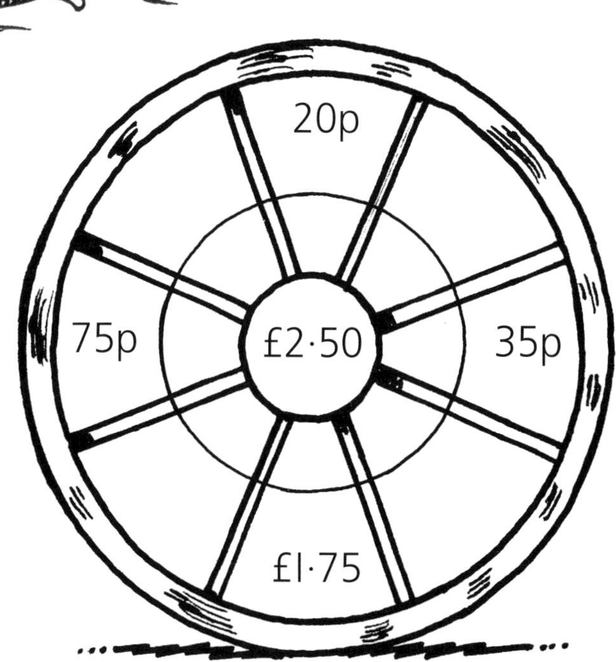

Fractions

Skills summary

- To recognise equivalence in fractions, e.g. $\frac{4}{8}$ and $\frac{2}{4}$
- To relate fractions to division, e.g. $\frac{1}{3}$ of 15 = 15 ÷ 3
- To compare and order fractions on a number line
- To reduce a fraction to its lowest terms by cancelling common factors
- To order fractions by converting them to fractions with a common denominator
- To order fractions by positioning them on a number line
- To add/subtract fractions by converting them to fractions with a common denominator

Core links

Number Textbook 1 page 26, top

- Write the fractions on the plants in order, smallest to largest.

Number Textbook 1 page 27, bottom

- Write ratios of different pairs in their simplest form, e.g. masks to non-masks.

Number Textbook 1 page 28, top

- Write each fraction as a percentage, to the nearest whole number.

Number Textbook 1 page 28, Explore

- Extend to five cards, using two cards for a 2-digit denominator.

Number Textbook 1 page 29, bottom

- Write each pair as a ratio in its simplest form.

Number Textbook 1 page 30, top

- Find the total of each set of three fractions.

Photocopy Master 17

- Extend to five cards, using two cards for a 2-digit denominator in one number.

Photocopy Master 18

- Convert each fraction to a decimal to the nearest hundredth, then to a percentage.

Photocopy Master 20

- Write the total and the difference between each pair of fractions.

Challenge Master 6

Extension Activities

- Investigate how many fraction pairs you can create in which one fraction has a denominator of 10. Convert the fractions into decimal form. Investigate pairs that have a fraction with a denominator of 20. Find an equivalent fraction with a denominator of 100, and convert these fractions into 2-place decimal numbers.
- Use number cards (0 to 9) to create pairs of the form $\frac{X}{X}$ = X·X, i.e. matching a fraction to a decimal, e.g. $\frac{1}{2}$ = 0·5.

Challenge Activity

- Using number cards (1 to 30), deal out two to make a fraction, placing the smaller on top. Reduce the fraction to its lowest terms. Repeat for different pairs of numbers.

Name ______________________________

Equivalent pairs

Use number cards 1 to 20.
Place four to make a pair of equivalent fractions.

The left fraction must be in its simplest form. For example:

Investigate how many different pairs of equivalent fractions you can make.

Addition

Skills summary

- To recall addition pairs to 10 and to 20
- To recall addition pairs to 100 (multiples of 10, multiples of 5)
- To recall addition pairs to 100 (any 2-digit number)
- To recall addition pairs to 1000 (multiples of 50, multiples of 10)
- To recall addition pairs to 1000 (any 3-digit number)
- To recall addition pairs to 1, e.g. 0·2 + 0·8, and to 10, e.g. 6·2 + 3·8
- To recognise what must be added to a 3-digit number to make the next multiple of 100
- To recognise what must be added to a 1-place decimal number to make the next whole number, next ten
- To recognise what must be added to a 2-place decimal number to make the next whole number, next ten
- To recognise what must be added to a 3-place decimal number to make the next whole number, next ten

Core links

Number Textbook 1 page 32, bottom

- Write each distance in feet, to the nearest foot (1 metre is approximately 3 feet).

Number Textbook 1 page 33, top

- Calculate what percentage the price of each ticket is of the most expensive, i.e. of the first one, at £9·36.

Number Textbook 1 page 34, top

- Write each distance in miles (8 kilometres is approximately 5 miles).

Number Textbook 1 page 35, top

- Write each capacity in pints, to the nearest pint (1 litre is $1\frac{3}{4}$ pints).

Photocopy Master 22

- Write each mileage in kilometres, as accurately as you can.

Photocopy Master 23

- Round each number to the nearest tenth and whole number.
- Invent a similar game, but include several numbers that have three decimal places.

Challenge Master 7

Extension Activity

- Round each time to the nearest hundredth of a second, tenth of a second and whole number of seconds.

Challenge Activity

- Using a set of number cards (0 to 9), deal out three to make a 3-digit number. Say the number that must be added to make 1000. Extend to dealing four cards to create a 3-place decimal number, and saying what must be added to make 10.

Name ______________________________

N10

Accurate times

This timer measures to a thousandth of a second.

For each time, write what must be added to reach the next whole number of seconds.

For example, 10·728 + 0·272 = 11 seconds

1

2

3

4

5

6

7

9

8

10

Addition and subtraction

Skills summary

- To mentally add several 1-digit numbers
- To mentally add two or three numbers (multiples of 10, multiples of 100)
- To mentally add several 2-digit numbers
- To use strategies to add several numbers, e.g. look for pairs to 10 (e.g. 7 + 3), and 100 (e.g. 70 + 30)
- To relate 'difference' to 'taking away'
- To mentally calculate difference by counting on from the smaller to the larger number
- To subtract using the strategy of mentally taking away by partitioning
- To add/subtract using the strategy of adding/subtracting a near multiple, then adjusting
- To choose an appropriate strategy for mental subtraction

Core links

Number Textbook 1 page 36, top

- Calculate the average (mean) amount spent per item in each question.

Number Textbook 1 page 36, bottom

- Calculate, as accurately as you can, the percentage of each person's savings the holiday costs.

Number Textbook 1 page 38, middle

- How much does each baby bear weigh in ounces (30 grams is approximately 1 ounce)?

Number Textbook 1 page 40, top

- Write each distance in miles.

Number Textbook 1 page 40, bottom

- Estimate, then calculate what percentage of the children, in each question, go home, and what percentage eat sandwiches.

Photocopy Master 24

- Calculate, for each grid, the total of each set of row totals, and the total of each set of column totals. Check that they are the same. Why are they the same?

Challenge Master 8

Extension Activities

- Investigate how many ways there are of making each of the numbers from 1 to 50 by adding consecutive numbers.
- Write a list of discoveries about adding consecutive numbers, e.g. 'all multiples of 3, except 3, can be written as the total of three consecutive numbers'.

Name ____________________________

N11 N12

Consecutive addition

18 can be made by adding consecutive numbers in two different ways.

18 = 3 + 4 + 5 + 6
18 = 5 + 6 + 7

Find two ways of making these numbers by adding consecutive numbers:

9	25	50	36
2 + 3 + 4			

Find three different ways of making these numbers by adding consecutive numbers:

21	15	33	60

Can you find four ways of making these numbers?

45	63	75

What other numbers can be made by adding consecutive numbers? Which numbers cannot be made? Can you see any patterns?

Multiples

Skills summary

- To recognise that a multiple of a number is the result of multiplying by the number
- To list the sequence of the first ten multiples of the numbers 2 to 10
- To list multiples of numbers beyond the tenth multiple
- To recognise a particular multiple of a number, e.g. the sixth multiple of 4
- To recognise multiples of numbers which are multiples of 10, e.g. multiples of 30, 50
- To recognise multiples which are common to two (or more) numbers
- To find the lowest common multiple of two numbers
- To find the lowest common multiple of a set of numbers

Core links

Number Textbook 1 page 42, top

- Write the next five numbers in each list, extending beyond 50.

Number Textbook 1 page 42, bottom

- For each number, list all the sets of multiples it belongs to.

Number Textbook 1 page 43, top

- Write the first three multiples of each set.

Number Textbook 1 page 43, bottom

- Find the total of each pair of fractions, expressed as simply as possible. Repeat for the difference.

Photocopy Master 26

- Extend each list to include a further five multiples.

Photocopy Master 28

- Make new grids by choosing your own headings, and recording the lowest common multiples.

Challenge Master 9

Extension Activities

- Investigate which numbers appear often in the table.
- Investigate any patterns in the table.
- Investigate patterns in the units digits of the numbers in the table.
- Investigate patterns in the digital roots (reduced numbers) of the numbers in the table.

Name ________________________________

N13

Common multiples

In each box, write the lowest common multiple of the row heading and the column heading.

	2	3	4	5	6	7	8	9	10	11
2			4							
3										
4						28				
5			20							
6		6								
7										
8										
9										
10										
11										

Division

Skills summary

- To divide by 2 by halving
- To divide by 4 by halving, then halving again
- To divide by 8 by halving, halving again, then halving again
- To divide by a multiple of 2, e.g. 6, by halving, then dividing by 3
- To recognise and use tests for divisibilty by 5, 10, 25 or 100
- To recognise and use tests for divisibilty by 2, 4, or 8
- To recognise and use tests for divisibilty by 3, 6 or 9
- To recognise and use tests for divisibilty by 7
- To recognise and use tests for divisibilty by 11

Core links

Number Textbook 1 page 45, top

- Which of these numbers are also exactly divisible by 8?

Number Textbook 1 page 45, top and bottom

- Are any of these numbers also divisible by 3, by 6, by 9?

Number Textbook 1 page 46, top

- For those numbers which are divisible by 3, write the result of dividing the number by 3. Repeat for those divisible by 6 and 9.

Number Textbook 1 page 46, top and bottom

- Are any of the numbers on this page divisible by 7?

Number Textbook 1 page 47, top

- Create a new table, writing a random arrangement of ten numbers in the left column, then complete it with ticks and crosses by checking for divisibility by different numbers.

Photocopy Master 29

- Test each number on the page for divisibility by other numbers.

Challenge Master 10

Extension Activity

- Explain why the test for divisibility by 7 works.

Challenge Activity

- Use a set of number cards (1 to 9). Shuffle them and deal out cards to show a 3-digit or 4-digit number. Score 1 point for divisibility by each of the numbers 2 to 9. Record the number and the points. Repeat for ten different numbers. Which number scored the most points? Which the least?

Name ______________________________

Divisibility by 7

Rachel says she has discovered a test for divisibility by 7, for numbers greater than 100.

For example, for 245, Rachel says:

double the hundreds digit	double 2 is 4
add this to the remaining part of the number	4 + 45 = 49
is this number divisible by 7?	

So, 245 is divisible by 7.

Now try Rachel's method on 725.

double the hundreds digit	double 7 is 14
add this to the remaining part of the number	14 + 25 = 39
is this number divisible by 7?	

So, 725 is not divisible by 7.

 Check Rachel's method to see if it works.

 Check these numbers for divisibility by 7.

1 483	**2** 257	**3** 468	**4** 728
5 371	**6** 645	**7** 539	**8** 294

 Can you discover a method which works for 4-digit numbers?

Inverses

Skills summary

- To recall multiplication facts up to 10 × 10
- To multiply TU × U by partitioning, mentally or using informal written methods
- To multiply HTU × U by partitioning, mentally or using informal written methods
- To multiply U·t × U by partitioning, mentally or using informal written methods
- To multiply U·th × U by partitioning, mentally or using informal written methods
- To use brackets to illustrate partitioning

Core links

Number Textbook 1 page 54, top

- Calculate how much is saved by taking a 15-mile journey with the cheapest taxi, compared with the other two.

Number Textbook 1 page 55, top

- Find the average cost per day for each holiday.

Number Textbook 1 page 57, top

- Write five of your own multiplications which have answers between 20 000 and 30 000.
- Investigate which is the closest answer you can find to 20 000 using a 4-digit number multiplied by a 1-digit number. You are not allowed to use the digit zero. Repeat for 25 000 and 30 000.

Number Textbook 1 page 59, top

- Write each area in square metres.
- Calculate the perimeter of each worktop.

Number Textbook 1 page 59, Explore

- Investigate how many different answers you can find between 15 000 and 25 000.

Photocopy Master 35

- Investigate, for each target, which number cards will give an answer as close to the target as possible.

Photocopy Master 39

- Create a similar game, but for multiplying 2-place decimals.

Challenge Master 11

Extension Activities

- Investigate other patterns formed when multiplying by 9.

Challenge Activity

- Use two sets of number cards (1 to 9). Shuffle them and deal out several sets of three cards to show a 2-digit number and a 1-digit number. Try each multiplication mentally, writing down only the answers. Check together. Extend to 3-digit numbers multiplied by 1-digit numbers, or 1-place and 2-place decimal numbers multiplied by 1-digit numbers.

Name ____________________

Multiplying nines

Try these multiplications.
Look for patterns.
Predict the next multiplication in each set.
Test your prediction by multiplying.

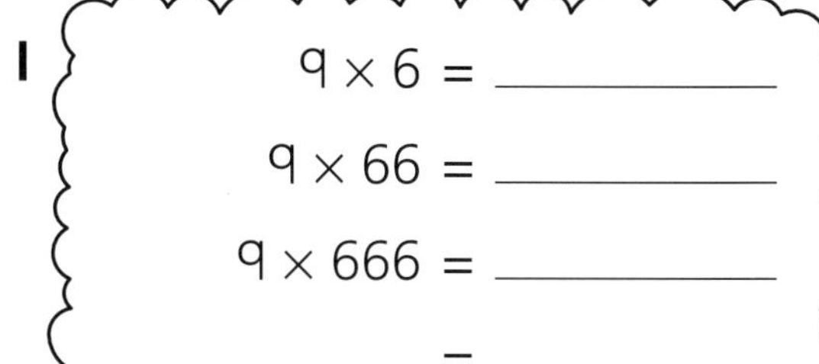

1

$9 \times 6 =$ ______

$9 \times 66 =$ ______

$9 \times 666 =$ ______

______ = ______

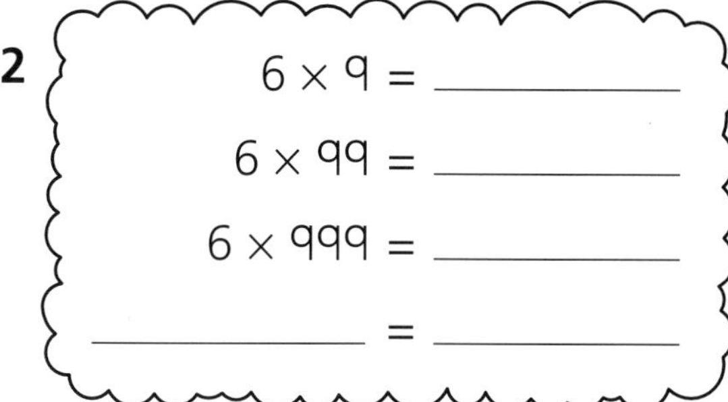

2

$6 \times 9 =$ ______

$6 \times 99 =$ ______

$6 \times 999 =$ ______

______ = ______

3

$9 \times 9 =$ ______

$99 \times 99 =$ ______

$999 \times 999 =$ ______

______ = ______

4

$6 \times 9 =$ ______

$66 \times 99 =$ ______

$666 \times 999 =$ ______

______ = ______

5

$12 \times 99 =$ ______

$23 \times 99 =$ ______

$34 \times 99 =$ ______

______ = ______

6

$9 \times 222222 =$ ______

$9 \times 333333 =$ ______

$9 \times 444444 =$ ______

______ = ______

7

$(9 \times 1) + 2 =$ ______

$(9 \times 12) + 3 =$ ______

$(9 \times 123) + 4 =$ ______

______ = ______

8

$(9 \times 9) + 7 =$ ______

$(9 \times 98) + 6 =$ ______

$(9 \times 987) + 5 =$ ______

______ = ______

Inverses

Skills summary

- To understand the relationship between addition and subtraction
- To recognise that from one multiplication or division fact, three others can be derived
- To understand the relationship between multiplication and division
- To check the result of an addition or subtraction by using the inverse operation
- To check the result of a multiplication or division by using the inverse operation

Core links

Number Textbook 1 page 60, top

- For each question, write some different multiplication facts which have the same answer, e.g. for 5×17, you could write $10 \times 8{\cdot}5$, $2{\cdot}5 \times 34$ etc.

Number Textbook 1 page 61, top

- Write your own set of six multiplications, then write a set of problems based on these.

Number Textbook 1 page 62, top

- Write some multiplication and division facts like these, then write some word problems based on them.

Number Textbook 2 page 57, top

- Write your own subtractions. Find as many different methods as you can of doing the subtractions, recording each method.

Challenge Master 12

Extension Activities

- Devise a similar trick using two dice, then three dice.
- Devise a similar trick using number cards (1 to 9).

Name ______________________________

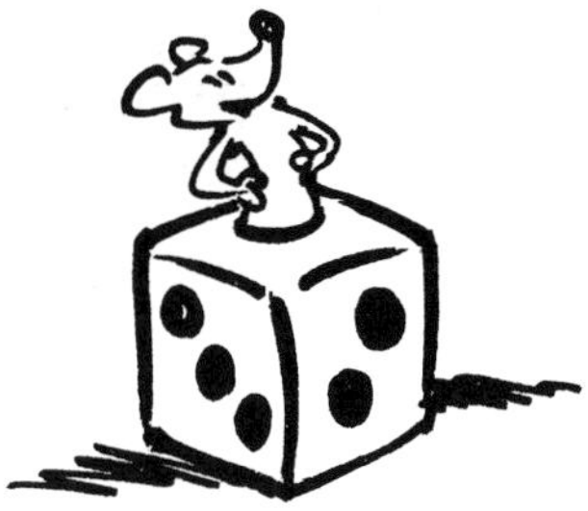

Dice trick

Throw three dice:

- double the first dice
- add 2
- multiply by 5
- add the second dice
- multiply by 10
- add the third dice
- subtract 100

 Try this several times.

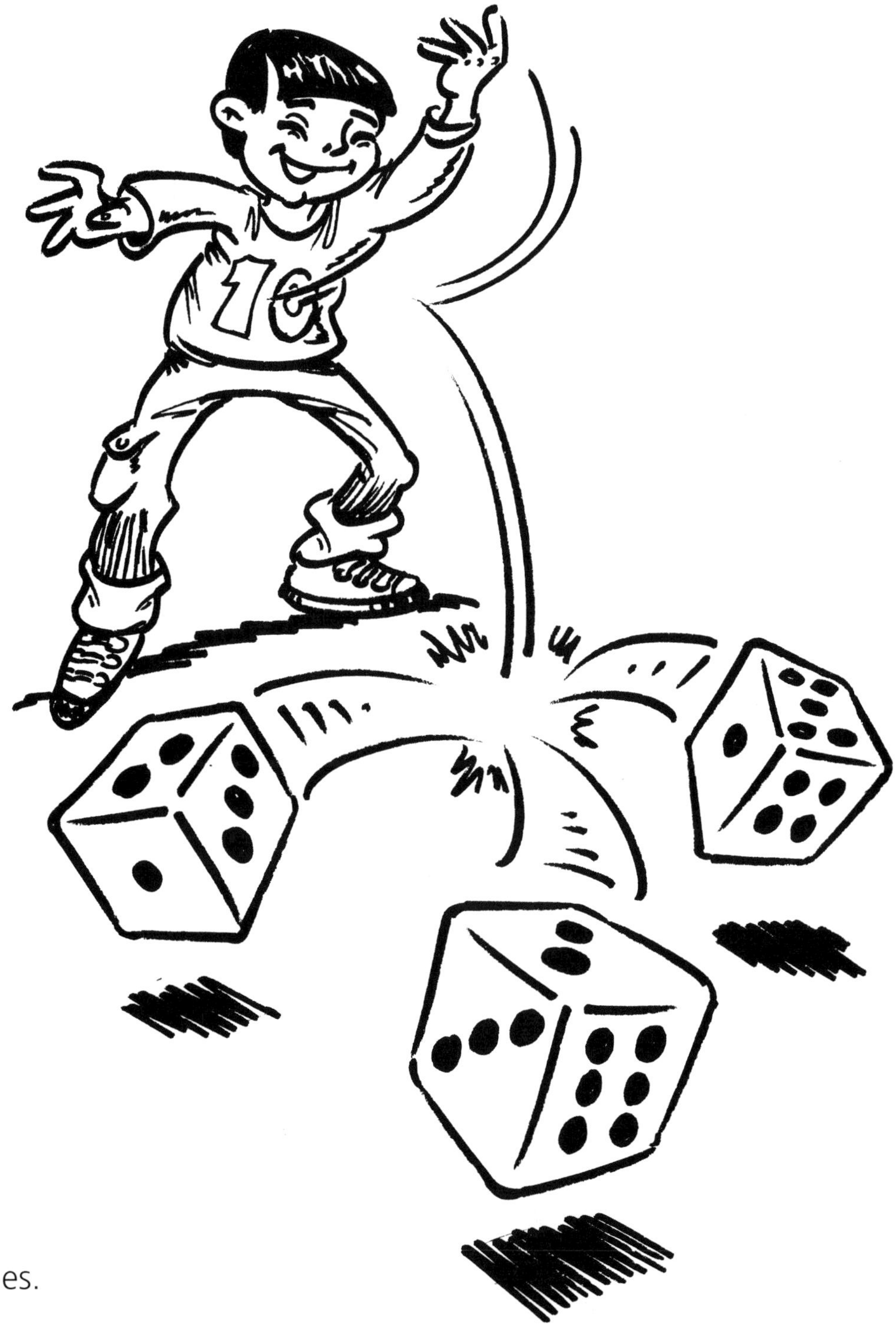

 Do you notice any pattern?

 Can you discover how the trick works?

Fractions

Skills summary

- To relate fractions to division, e.g. $15 \div 3 = \frac{1}{3}$ of 15
- To convert a mixed number into an improper fraction, and vice versa
- To record a remainder as a fraction
- To record a remainder as a fraction in its lowest terms
- To record a remainder, when dividing by 10, as a fraction and as a decimal, e.g. $63 \div 10 = 6\frac{3}{10}$ or 6·3
- To interpret remainders as fractions, and record them in decimal notation, e.g. $57 \div 4 = 14\frac{1}{4} = 14{\cdot}25$
- To write a quotient as a decimal, rounding to one decimal place, e.g. $29 \div 7 = 4{\cdot}1$

Core links

Number Textbook 1 page 63, top and bottom

- Write each remainder as a decimal to the nearest tenth.

Number Textbook 1 page 65, top

- Write each amount in pints, to the nearest pint (1 litre is approximately $1\frac{3}{4}$ pints). Convert each answer to gallons and pints.

Number Textbook 1 page 66, top and bottom

- Write each mixed number as a decimal to the nearest tenth.

Number Textbook 1 page 68, top

- Write each mixed number as a decimal to the nearest tenth.

Photocopy Master 42

- Play the game, but instead of just giving the remainder, give the answer, e.g. if you remove 25 and throw a 6, you say 'four and one sixth'.

Photocopy Master 43

- Write each answer in decimal form, correct to one decimal place.

Challenge Master 13

Extension Activities

- Investigate other similar possible fractions that are equivalent to a 1-place decimal.
- Extend the activity to using five numbers which link a fraction to a 2-place decimal.
- Using number cards (1 to 20), investigate how many 1-place decimal and fraction equivalents can be found. How many can you find involving 2-place decimals?

Challenge Activity

- Use a set of number cards (1 to 9). Shuffle them and deal out two cards to show a fraction – the first number is the numerator, the second is the denominator. If it is an improper fraction, first convert it into a mixed number. Express the fraction as a decimal correct to two decimal places.

Name ______________________________

Fractions and decimals

Copy the four numbers into the boxes, to make a fraction equivalent to a decimal.

1 (3, 0, 5, 6) $\frac{3}{5}$ = 0 · 6

2 (5, 1, 6, 2) $\frac{\square}{\square}$ = □ · □

3 (4, 4, 10, 0) $\frac{\square}{\square}$ = □ · □

4 (7, 5, 1, 4) $\frac{\square}{\square}$ = □ · □

5 (23, 5, 4, 6) $\frac{\square}{\square}$ = □ · □

6 (1, 10, 12, 2) $\frac{\square}{\square}$ = □ · □

7 (9, 1, 8, 5) $\frac{\square}{\square}$ = □ · □

8 (1, 5, 8, 6) $\frac{\square}{\square}$ = □ · □

9 (17, 2, 8, 5) $\frac{\square}{\square}$ = □ · □

10 (44, 5, 8, 5) $\frac{\square}{\square}$ = □ · □

Decimals

Skills summary

- To recognise place-value in numbers with more than four digits
- To recognise place-value in numbers with up to three decimal places
- To know the value of each digit in a number with up to three decimal places
- To round a 1-place decimal number to its nearest whole number
- To round a 2-place decimal number to its nearest whole number, tenth
- To round a 3-place decimal number to its nearest whole number, tenth, hundredth
- To order a set of decimal numbers with up to three decimal places
- To relate metres and kilometres using 3-place decimals
- To relate litres and millilitres using 3-place decimals
- To convert a fraction to a decimal to the nearest three decimal places

Core links

Number Textbook 1 page 69, top

- Round each decimal to the nearest whole number, tenth, hundredth and thousandth.

Number Textbook 1 page 69, bottom

- Write each distance in inches (12 inches is approximately 30 cm).

Number Textbook 1 page 70, bottom

- Write each weight in pounds (2·2 pounds is approximately 1 kilogram).

Photocopy Master 45

- Round each decimal number to its nearest tenth and nearest whole number.

Photocopy Master 46

- Make 3-place decimal numbers with the cards, and round them to the nearest tenth and hundredth.

Photocopy Master 47

- Write the difference between each pair, both as a fraction and as a decimal.

Challenge Master 14

Extension Activities

- Investigate which denominators produce recurring decimals.
- Investigate which fractions produce a decimal with one recurring digit and which produce decimals with a recurring string of digits.
- Investigate how many different fractions that produce recurring decimals can be created using the digits 0 to 9.

Challenge Activity

- Using a set of number cards (1 to 10), deal out five. Explore how many different fractions equivalent to recurring decimals can be created using these five numbers, e.g. 2, 3, 4, 5 and 8 leads to $\frac{2}{3}$, $\frac{4}{3}$, $\frac{5}{3}$, $\frac{8}{3}$.

Name ________________________________

Recurring decimals

To convert a fraction to a decimal, you can use a calculator. For example, to find $\frac{3}{8}$ as a decimal divide 3 by 8.
On the calculator press:

 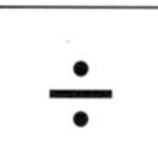  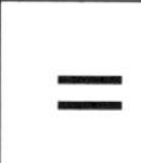

3 ÷ 8 =

The calculator shows this: 0·375.

To convert $\frac{2}{3}$ to a decimal divide 2 by 3.
On the calculator press:

 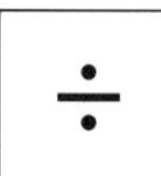  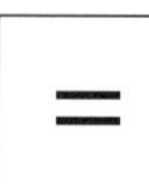

2 ÷ 3 =

The calculator shows this: 0·666666.
This is called a **recurring decimal**.
The digit 6 recurs.

Investigate converting different fractions to decimals.

Which have 1 decimal place?
Which have 2 decimal places?
Which have 3 decimal places?
Which are recurring decimals?

Addition

Skills summary

- To add TU + TU using informal and standard written methods
- To add HTU + TU and HTU + HTU using informal and standard written methods
- To add ThHTU + ThHTU using standard written methods
- To add U·t + U·t using informal and standard written methods
- To add U·th + U·t and U·th + U·th using standard written methods
- To add a set of whole numbers and decimal numbers with mixed numbers of digits and mixed numbers of decimal places
- To estimate the answer to additions, by rounding, before calculating

Core links

Number Textbook 2 page 4, top

- Convert each distance into miles (50 miles is approximately 80 kilometres).

Number Textbook 2 page 6, top

- Compare the cost of each item to the most expensive, i.e. the speed boat.
- Find the total cost of all nine items, then write each as a percentage of the total cost.

Number Textbook 2 page 7, top

- Measure the lengths of the sides of the classroom in metres, and calculate its perimeter. Find the perimeter of some other parts of the school.

Number Textbook 2 page 9, top

- Try estimating each time by starting and stopping a stopwatch, without looking. Check to see how close your estimate is.

Number Textbook 2 page 10, top

- Write each answer in miles, as accurately as you can (50 miles is approximately 80 kilometres).

Photocopy Master 48

- Create a variation of the game, adding decimal numbers.

Photocopy Master 52

- Investigate what happens if you change the order of the numbers on the bottom layer in question 1. Which order gives the largest/smallest top number? Investigate ways of predicting the top number, by looking only at the bottom four numbers.

Challenge Master 15

Extension Activities

- Invent your own 'Spot the mistakes' sheet. Include four errors and ten correct answers. Ask a friend to spot the mistakes.

Challenge Activity

- Use a set of number cards (0 to 9), and counters for decimal points. Deal the cards out in threes to create three 2-place decimal numbers. Mentally try to calculate the total. Write down your answer. Now find the total using the standard written method. Was your mental calculation correct?

Name ____________________

N23 N24

Spot the mistakes

a	b	c	d
4·95	1·7	16	35·8

e	f	g	h
2·736	0·048	0·9	5·496

Four of these additions are wrong.
Can you spot the mistakes, and write the correct answers?

1 a + b = 6·65

2 e + g = 3·636

3 f + e = 2·774

4 c + d = 51·8

5 h + e = 8·232

6 a + d = 40·75

7 b + c + g = 18·6

8 c + f + d = 51·848

9 a + h + f = 9·494

10 a + b + h = 12·146

11 c + b + d + a = 58·45

12 e + f + c + a = 23·834

13 g + h + d + c = 58·196

14 h + c + e + b = 26·932

Write down which four of the numbers add to give these numbers.

15 12·194

16 20·484

17 48·982

Subtraction

Skills summary

- To subtract TU – TU using informal and standard written methods
- To subtract HTU – TU and HTU – HTU using informal and standard written methods
- To subtract ThHTU + ThHTU using standard written methods
- To subtract U·t – U·t using informal and standard written methods
- To subtract U·th – U·t and U·th – U·th using standard written methods
- To subtract one number from another, each with mixed numbers of digits and mixed numbers of decimal places
- To estimate the answer to subtractions, by rounding, before calculating

Core links

Number Textbook 2 page 13, top

- Estimate, then calculate, the approximate percentage of each crowd which is adults and which is children.
- Collect some football attendances from the newspapers. Calculate the differences between the attendances at pairs of clubs.

Number Textbook 2 page 14, top

- Calculate, in each question, what percentage of the savings are being spent.

Number Textbook 2 page 15, bottom

- Write each answer to the nearest pint, as accurately as you can (1 litre is approximately $1\frac{3}{4}$ pints).

Number Textbook 2 page 16, top

- If the world record is 19·35 seconds, by how much does each runner need to improve to match it?

Number Textbook 2 page 17, top

- Calculate what percentage of each amount is spent.

Photocopy Master 57

- Extend the game to a subtraction involving 3-place decimal numbers.

Photocopy Master 58

- Round each answer to its nearest whole number and tenth.

Challenge Master 16

Extension Activity

- Choosing the first question, for example, investigate changes in the top number when the bottom numbers are in a different order.

Challenge Activity

- Use a set of number cards (0 to 9), and counters for decimal points. Deal the cards out in threes to create two 2-place decimal numbers. Mentally try to subtract the smaller from the larger. Write down your answer. Now subtract using the standard written method. Was your mental calculation correct? Repeat the activity several times.

Name ________________________________

Difference pyramids

The number in each block is the difference between the numbers in the two blocks directly below.

Complete the pyramids below by writing a number in each block.

1

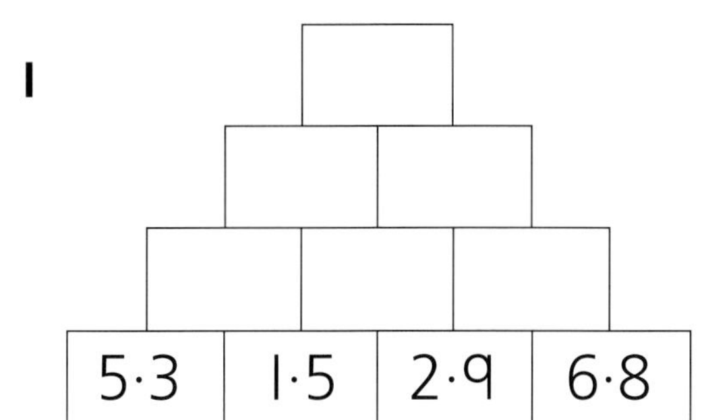

2

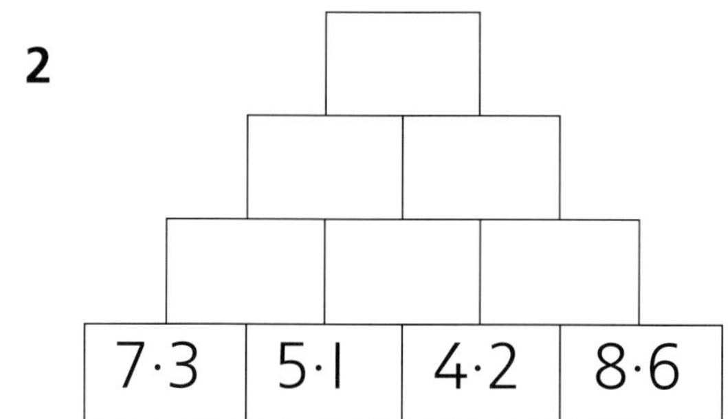

3

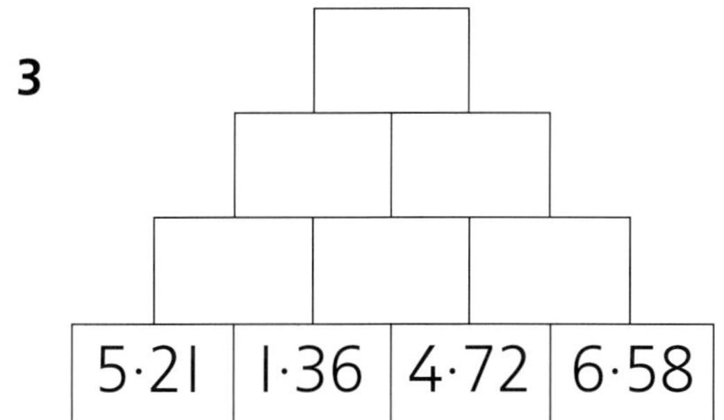

4

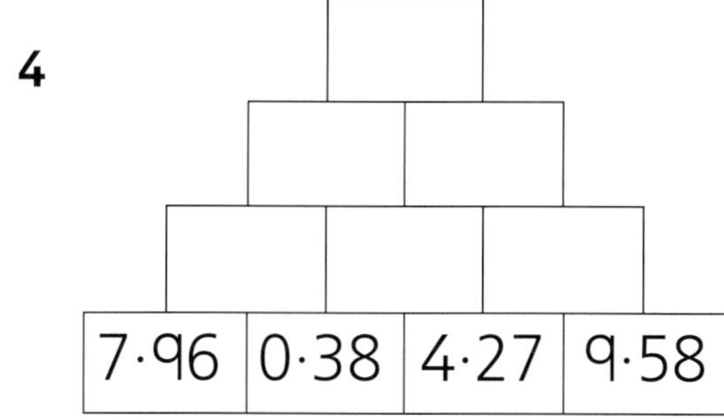

5

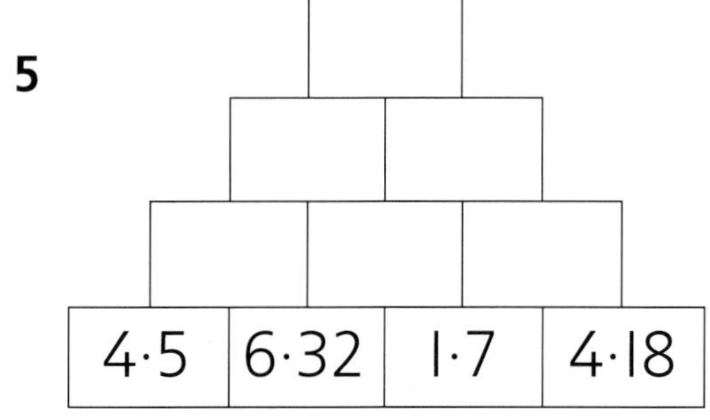

6

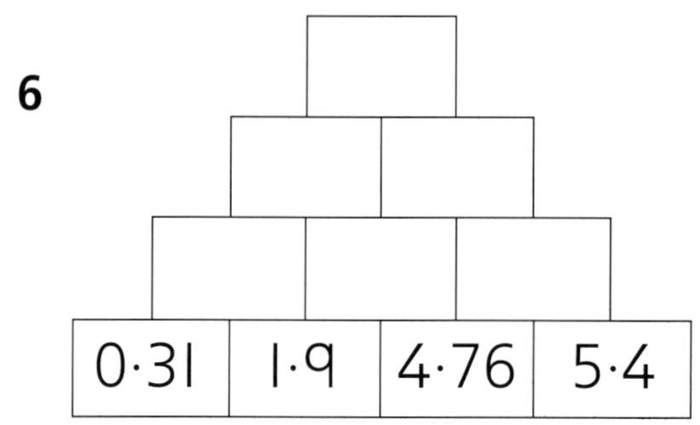

Create your own difference pyramids.
Include at least one 3-place decimal.

Factors and primes

Skills summary

- To recognise a pair of factors of a number
- To recognise that any number has itself and 1 as factors
- To find all possible factor pairs for a number
- To list the factors of a number
- To recognise prime numbers as numbers which have exactly two factors
- To recognise the prime factors of numbers
- To reduce a number to its set of prime factors
- To recognise square numbers as numbers which have an odd number of factors
- To recall immediately, whether or not one number is a factor of another
- To find common factors of two numbers
- To recall prime numbers between 1 and 30
- To recall prime numbers between 1 and 100

Core links

Number Textbook 2 page 19, top and bottom

- Reduce each number to its set of prime numbers.

Number Textbook 2 page 20, top

- List all the factors of each number.

Number Textbook 2 page 21, top

- Explain why the square numbers have an odd number of factors. Hint: list the factor pairs of the square numbers.

Number Textbook 2 page 23, top

- Check that multiplying together the circled prime numbers leads back to each original number.

Photocopy Master 59

- Highlight six numbers on the board which you think have the most factors, and six which you think have the fewest factors. Investigate the number of factors of each number on the board, then check your original estimates.
- How many different possible dice totals are factors of each number on the board?

Photocopy Master 60

- Investigate patterns in the completed table, e.g. patterns in the 4s, the 7s etc.
- Complete the totals of factors of each number by adding along each row, and writing the total beyond the right-hand column.

Photocopy Master 62

- Extend to making numbers other than prime numbers, e.g. square numbers.

Challenge Master 17

Extension Activity

- Can you find another set of four digits which will complete all the sentences?

Challenge Activity

- Use a set of number cards (0 to 9). Deal out five cards. Investigate how many different 2-digit prime numbers can be created with pairs of these cards.

Name ______________________________

Factor sentences

Use these four number cards: 2 4

Each digit can be used up to four times.
How many correct sentences can you complete?

2 is a factor of 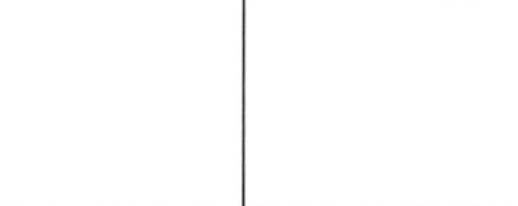

3 is a factor of 

4 is a factor of 

5 is a factor of 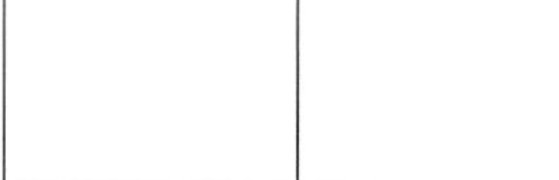

6 is a factor of 

7 is a factor of 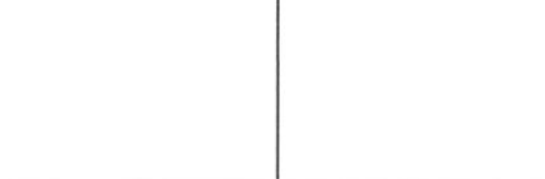

8 is a factor of 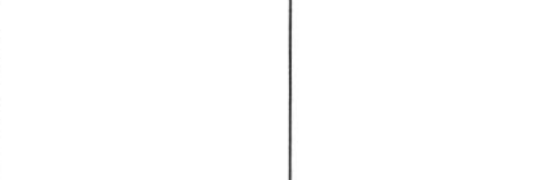

9 is a factor of 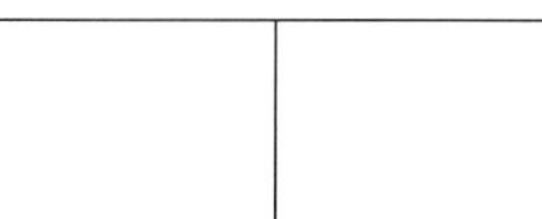

 Try with these number cards:

 Try with your own set of digits.

Number sequences

Skills summary

- To locate the position of positive and negative whole numbers (integers) on a number line
- To order a pair of positive and negative numbers
- To order a set of positive and negative numbers
- To find the difference between two numbers, positive or negative
- To add a pair of positive and negative numbers
- To add a set of positive and negative numbers
- To subtract one number from another when each is either positive or negative
- To recognise patterns in number sequences, e.g. triangular numbers
- To construct different number sequences

Core links

Number Textbook 2 page 25, top

- Write the total of each set of three numbers.

Number Textbook 2 page 26, top

- Explore sets of translations which take you from one point on the grid to another, e.g. from the diver to the submarine (vertically $^{-}4$, horizontally $^{+}9$).

Number Textbook 2 page 65, bottom

- Explore patterns in the list of triangular numbers, e.g. differences between them, sums of consecutive pairs etc.

Number Textbook 2 page 66, Explore

- Start with a number, say 58, and try to find a pair of starting numbers which ensure that 58 is in the sequence, e.g. 1, 4.

Number Textbook 2 page 67, Explore

- Investigate patterns in the sequence of cubic numbers, e.g. the units digits are: 1, 8, 7, ...

Photocopy Master 65

- Change the tables from difference tables to addition tables.

Photocopy Master 90

- Explore different ways of obtaining 0 heads, 1 head and 2 heads when tossing two coins. Extend to tossing three coins. See how this relates to Pascal's triangle. Check for tossing four coins.

Challenge Master 18

Extension Activities

- Investigate patterns in the units digits of a sequence of triangular numbers.
- Construct Pascal's triangle, and look for different sequences within it, including the triangular numbers.
- Investigate any pattern in the position of triangular numbers on a multiplication square.

Name ______________________________

Triangular numbers

Write the first 12 triangular numbers.

1 3 6 10 15 21 ______________________________

Look at sets of three consecutive triangular numbers.

For example, 3 6 10

square the middle number	$6 \times 6 = 36$
multiply the outside numbers	$3 \times 10 = 30$
find the difference	$36 - 30 = 6$

Try with these numbers: 10, 15, 21.

square the middle number	$15 \times 15 = 225$
multiply the outside numbers	$10 \times 21 = 210$
find the difference	$225 - 210 = 15$

Investigate for other sets of three triangular numbers.
What pattern do you notice?

Write the squares of the triangular numbers:

1 9 36 ______________________________

Look at pairs of consecutive squares of triangular numbers.

consecutive squares	1, 9
find the difference	8

Investigate for other pairs. What patterns do you notice?

Multiplication

Skills summary

- To multiply a 1-digit number by a multiple of 10 or 100, e.g. 7 × 30, 6 × 400
- To multiply a 2-digit number by 10, by 100, e.g. 35 × 10, 47 × 100
- To multiply TU × U by partitioning, e.g. 5 × 16 = (5 × 10) + (5 × 6), using informal and standard written methods
- To multiply HTU × U using informal and standard written methods
- To multiply TU × TU using informal and standard written methods
- To multiply ThHTU × U using informal and standard written methods
- To multiply HTU × TU using standard written methods
- To multiply U·t × U using informal and standard written methods
- To multiply U·th × U using standard written methods
- To multiply U·th × TU using standard written methods
- To multiply U·t × U·t using standard written methods

Core links

Number Textbook 2 page 31, top

- Measure accurately the width of some boxes in centimetres and millimetres. Record these in centimetres in decimal form, then calculate the total width for different numbers of each box.

Number Textbook 2 page 32, top

- Write the height of each building to the nearest foot (1 metre is approximately 3 feet).

Number Textbook 2 page 34, top

- Write each time, as accurately as you can, as a percentage of a minute.

Number Textbook 2 page 36, Explore

- Extend to using five cards and multiplying a 3-place decimal number.

Photocopy Master 68

- Repeat the activity, but using five cards and multiplying a 3-place decimal number.

Photocopy Master 67

- Create a new sheet involving multiplying 2-place and 3-place decimals. Complete the sheet by rolling a dice, then multiplying.

Challenge Master 19

Extension Activities

- Use the Italian method to multiply some 3-digit numbers by 2-digit and 3-digit numbers. Use a calculator to check the answers.
- Use the Italian method to multiply two 1-place decimal numbers. Extend to a mixture of 1-place and 2-place decimals. Use a calculator to check the answers. Explore rules for locating the position of the decimal point in the answer.

Name ______________________________

N31 N32

Italian multiplication

Use this multiplication method to multiply 54 × 26.

Draw a 2 × 2 square grid with diagonals in each square.
Write the two numbers in position, as shown.

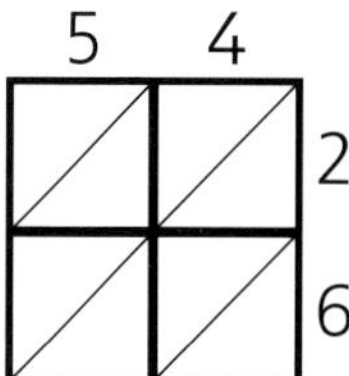

Multiply each row heading by each column heading.
Write the tens digit of the answer above the diagonal, and the units digit below.

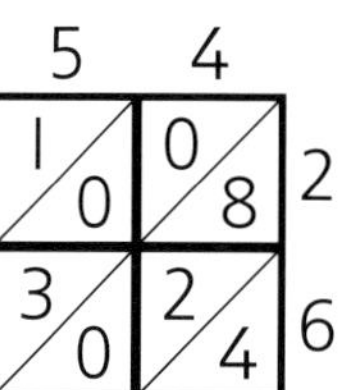

Add the numbers in each diagonal row of the grid, starting from the bottom row. Carry numbers to the next diagonal up if necessary.

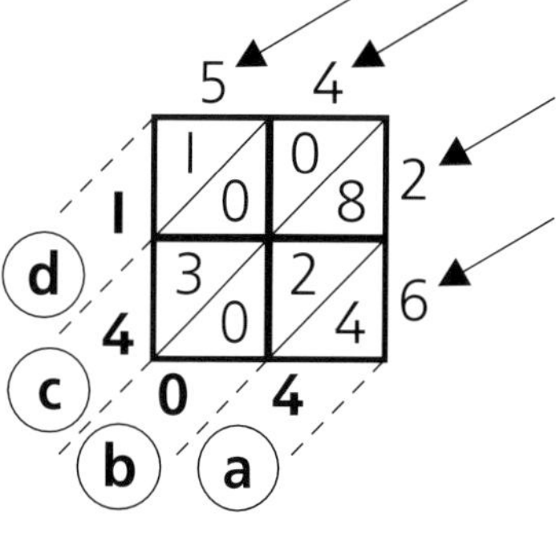

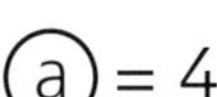

(a) = 4
(b) = 8 + 2 + 0 = 10 (carry 1 to diagonal above)
(c) = 0 + 0 + 3 + 1 = 4
(d) = 1

Write the answer.

54 × 26 = 1404

Use the Italian method of multiplication to multiply these numbers.
Check each answer with a calculator.

1 34 × 28 **2** 42 × 17 **3** 45 × 29
4 19 × 36 **5** 85 × 24 **6** 73 × 36

Try these, using a 2 × 3 grid.

7 142 × 16 **8** 351 × 17 **9** 468 × 29
10 358 × 42 **11** 417 × 65 **12** 732 × 48

Division

Skills summary

- To divide by grouping into sets
- To use lists of multiples to recognise division facts
- To recognise and record remainders when dividing
- To decide whether to round up or round down after division
- To divide TU ÷ U using informal written methods
- To divide TU ÷ U using standard written methods (short division)
- To divide HTU ÷ U using informal written methods
- To divide HTU ÷ U using standard written methods (short division)
- To divide HTU ÷ TU using standard written methods (long division)
- To divide TU·t ÷ U and TU·th ÷ U using standard written methods
- To divide one decimal number by another by first multiplying both by 10 or 100

Core links

Number Textbook 2 page 37

- Calculate the total amount spent and the total number of tickets bought. Estimate the average (mean) amount spent per ticket. Use a calculator to check the answer and compare it with your estimate.

Number Textbook 2 page 38, bottom

- Calculate the total number of people transported and the total number of buses used. Estimate the average (mean) number of people per bus. Use a calculator to check the answer and compare it with your estimate.

Number Textbook 2 page 40, top

- Find the average time per day each person spent not working over the five days.
- Find the average time spent working/not working over 7 days.

Number Textbook 2 page 40, bottom

- Write the mean weight of each fruit to the nearest pound (1 kilogram is approximately 2·2 pounds).

Photocopy Master 71

- Collect the attendance figures of football matches from the Sunday paper. Calculate the mean attendance for each division.

Challenge Master 20

Extension Activities

- Write ten divisions like the last nine on the sheet. Write them on cards, one per card. Estimate the order of size of the answers by placing the cards in a line from smallest estimated answer to largest. Divide each by using the 'multiplying by 10 or 100' trick, then check your estimated order.

Name ___________________________

Making division easier

Some divisions can be made easier by first dividing both numbers by 10 or 100.

$40 \div 30 = 4 \div 3$ (dividing both by 10)

or

$840 \div 700 = 8{\cdot}4 \div 7$ (dividing both by 100)

 Try these:

1 $74 \div 20$ **2** $315 \div 50$ **3** $91 \div 70$

4 $720 \div 300$ **5** $126 \div 60$ **6** $760 \div 200$

7 $240 \div 400$ **8** $640 \div 40$ **9** $8500 \div 500$

Some divisions can be made easier by first multiplying both numbers by 10 or 100.

$0{\cdot}45 \div 0{\cdot}9 = 4{\cdot}5 \div 9$ (multiplying both by 10)

or

$1{\cdot}28 \div 0{\cdot}08 = 128 \div 8$ (multiplying both by 100)

 Try these:

10 $1{\cdot}2 \div 0{\cdot}3$ **11** $11{\cdot}5 \div 0{\cdot}5$ **12** $23{\cdot}4 \div 0{\cdot}9$

13 $1{\cdot}32 \div 0{\cdot}02$ **14** $0{\cdot}84 \div 0{\cdot}7$ **15** $2{\cdot}16 \div 0{\cdot}06$

16 $0{\cdot}81 \div 0{\cdot}09$ **17** $0{\cdot}96 \div 0{\cdot}4$ **18** $0{\cdot}96 \div 0{\cdot}08$

Percentages, fractions and decimals

Skills summary

- To record fractions (tenths and hundredths) using decimal notation
- To record 1- and 2-place decimals using fraction notation
- To convert simple fractions ($\frac{1}{2}$, $\frac{1}{4}$, $\frac{3}{4}$) to decimals
- ⬇ To convert other fractions, e.g. $\frac{4}{5}$, $\frac{17}{25}$, to decimals
- To recognise a percentage as a number of hundredths
- To record a fraction (tenths or hundredths) as a percentage
- To recognise the percentage represented by simple fractions ($\frac{1}{2}$, $\frac{1}{4}$, $\frac{3}{4}$)
- To convert simple fractions to percentages, and vice versa
- ⬇ To convert any fraction to a percentage
- ⬇ To relate a percentage to its matching fraction and decimal number
- ⬇ To express a percentage as a fraction in its lowest terms.
- To find simple percentages of amounts
- ⬇ To increase and decrease an amount by a given percentage

Core links

Number Textbook 2 page 44, top
⬇ Calculate the percentage of the total number of children who voted 'yes'.

Number Textbook 2 page 45, top
⬇ Calculate the percentage of the total number of filmgoers who are children.

Number Textbook 2 page 46, top
⬇ Find what percentage of all the children in both schools come from each town.

Number Textbook 2 page 48, middle
- Write the difference between each pair as a percentage.

Number Textbook 2 page 48, bottom
- Write the difference between pairs in each set, as fractions in their lowest terms.

Number Textbook 2 page 49, top
- Write the fractions in order, smallest to largest.

Photocopy Master 75
- Write each percentage as a fraction in its lowest terms.

Photocopy Master 76
- Record 20, 25 or 50 sets of football results, then find percentages related to them.

Photocopy Master 78
⬇ Extend to 2-place and 3-place decimals, e.g. $\frac{3}{4}$ = 0·75, $\frac{1}{8}$ = 0·375.

Challenge Master 21

Extension Activities
⬇ Use a catalogue. Calculate new prices for different percentages discounted.

Challenge Activity
- Use a set of number cards (1 to 10). Deal them out in pairs and create a fraction with each pair, placing the smaller number on top. Write the five fractions down, then convert each to a decimal.

Name ______________________

N35 N36

Sales

The shop is holding a sale.
Reduce the prices.

TOY SHOP
SALE
PRICES SLASHED

20% off

1 £3·20

sale price

2 £4·60

sale price

3 £5·80

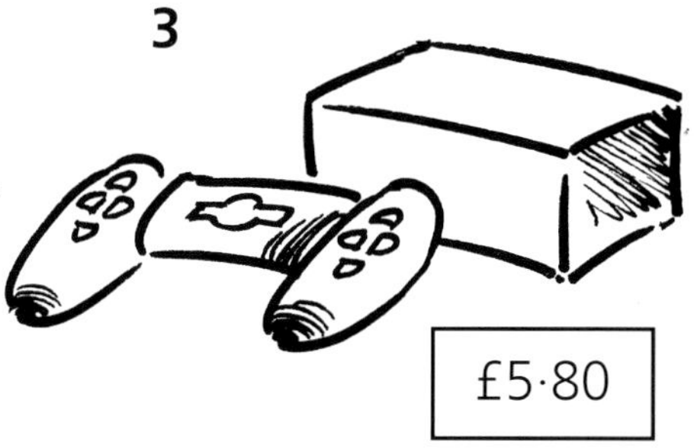

sale price

30% off

4 £7·60

sale price

5 £9·40

sale price

6 £4·40

sale price

40% off

7 £8·40

sale price

8 £5·20

sale price

9 £6·60

sale price

Write the sale prices if everything is reduced by 15%.

Ratio and proportion

Skills summary

- To compare two amounts by expressing them as a ratio
- To compare two amounts by expressing them as a ratio in its simplest form
- To compare more than two amounts by expressing them as a ratio in its simplest form
- Given a total amount divided into a given ratio, calculate the value of each part
- Express a proportion of a part to a whole as a fraction, a decimal, a percentage
- Express a proportion of a part to a whole in its simplest form
- Solve problems involving proportion
- Solve problems involving ratio

Core links

Number Textbook 2 page 50, top
- Write the percentage of each grid which is coloured and which is not coloured.

Number Textbook 2 page 50, bottom
- Write each proportion as a percentage.

Number Textbook 2 page 51, top
- Write each sleeping time as a percentage of the day.

Number Textbook 2 page 52, top
- Collect weather data over a month, and calculate similar proportions.

Number Textbook 2 page 53, top
- Calculate the percentages of each strip which are coloured in each colour.

Number Textbook 2 page 55, Explore
- Extend to different numbers of cubes, e.g. 36, 48.

Photocopy Master 82
- Write each proportion as a percentage.
- Collect real data from the newspapers, and write proportions and percentages.

Photocopy Master 84
- Write each of the shaded and non-shaded areas as a percentage of each grid.

Challenge Master 22

Extension Activity
- Express each shaded part as a proportion of the whole grid as a fraction in its lowest terms, a decimal and a percentage.

Challenge Activity
- Use a set of number cards (1 to 20). Deal them out in pairs. Write each pair as a ratio in its simplest form. Extend to dealing out the cards in threes and repeating.

Name ______________________________

N37 N38

Grid ratios

Draw a rectangular grid.

Shade some squares, and leave some blank.

Explore ways of shading so that you make different ratios of shaded to blank of the form 1:X. Look at these examples.

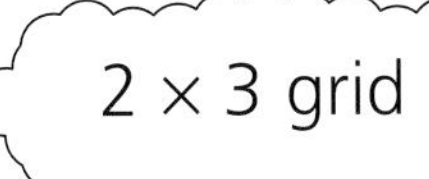

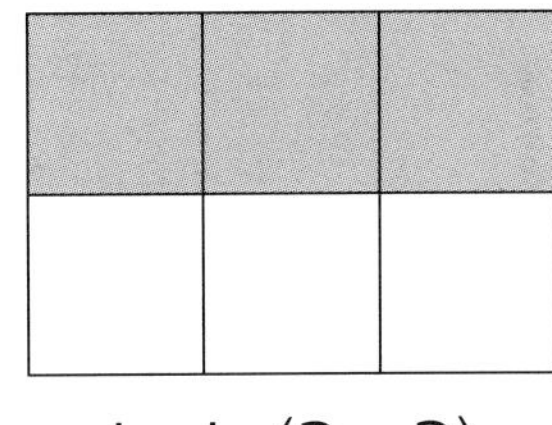

1 : 1 (3 : 3)

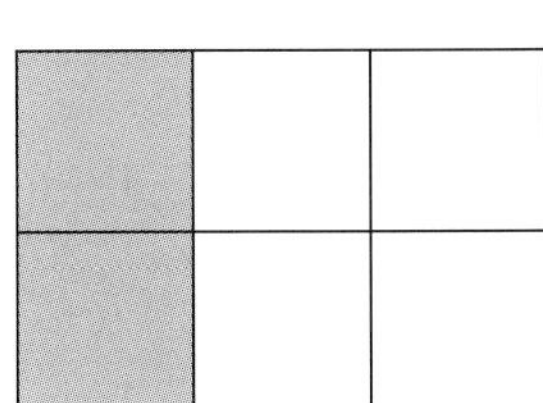

1 : 2 (2 : 4)

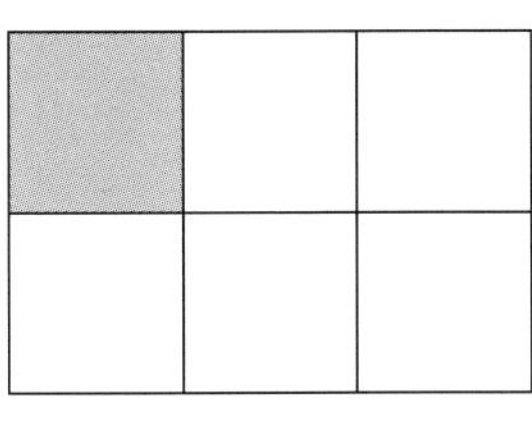

1 : 5

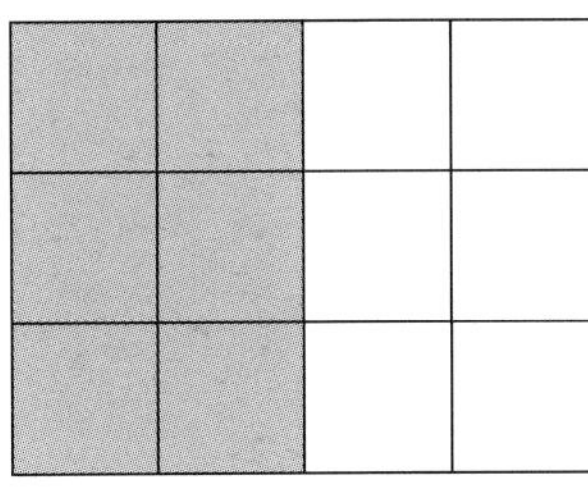

1 : 1 (6 : 6)

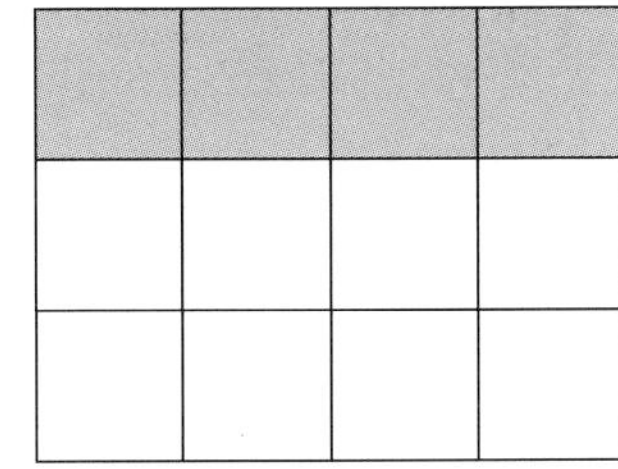

1 : 2 (4 : 8)

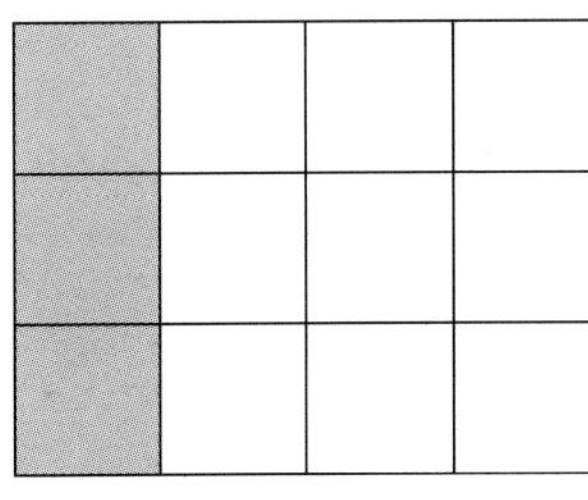

1 : 3 (3 : 9)

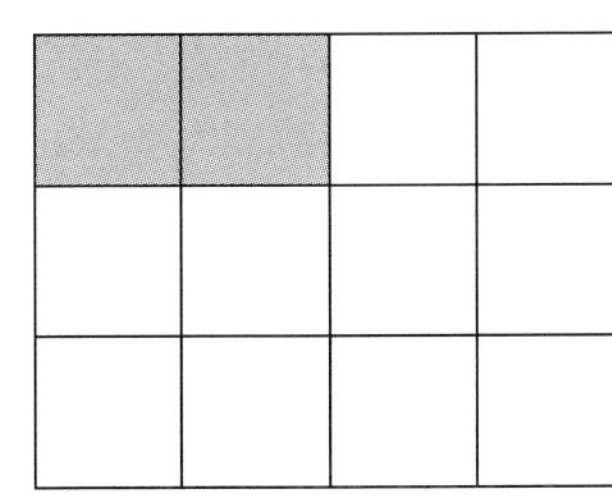

1 : 5 (2 : 10)

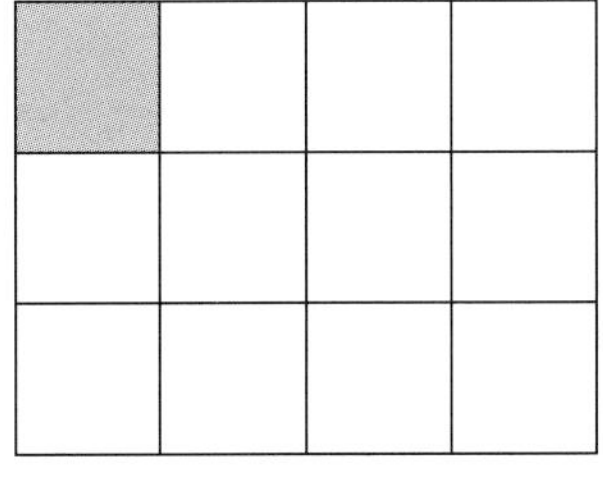

1 : 11

Investigate for different-sized grids.

Number patterns

Skills summary

- To recognise odd and even numbers up to 10
- To recognise odd and even numbers up to 100
- To recognise larger numbers as odd or even
- To recognise pattern in the oddness or eveness of the answers when adding odd/even numbers
- To recognise pattern in the oddness or eveness of the answers when subtracting odd/even numbers
- To recognise pattern in the oddness or eveness of the answers when multiplying and dividing odd/even numbers
- To calculate the reduced number (the digital root) of a number
- To recognise patterns in the digits of arrangements of numbers and number sequences

Core links

Number Textbook 2 page 60, Explore

- For any chosen set of three cards, investigate how many different numbers can be made, e.g. with 3, 4 and 7, you can make 14 = 3 + 4 + 7, 25 = (4 × 7) – 3 etc.
- Extend to choosing four cards and trying to make the numbers from 1 to 30.

Number Textbook 2 page 60, top

- Use ten points equally spaced round a circle. Number them 0 to 9 in sequence. Draw patterns representing the number sequences by drawing a straight line from the first number to the second, then the second number to the third, and so on ... How do the patterns compare?

Number Textbook 2 page 69, top

- Express the proportion of odd and even numbers as percentages of the whole grid.

Number Textbook 2 page 70, Explore

- Write comments related to any symmetrical patterns in the grid. Look for a centre of rotational symmetry on part of the grid.

Challenge Master 23

Extension Activities

- Investigate patterns in the digit products of the numbers in the multiplication table. To find the digit product, keep multiplying the digits until you reach a 1-digit number, e.g. 18 ⇒ 8, 39 ⇒ 27 ⇒ 14 ⇒ 4.

Name ___________________________

N40 N43

Reduced number patterns

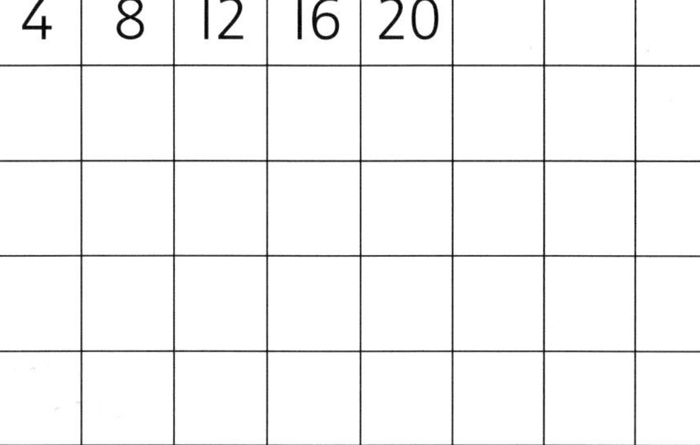

Complete this multiplication table.

1	2	3	4	5	6	7	8
2	4	6	8	10	12	14	16
3	6	9	12	15	18	21	24
4	8	12	16	20			

To find the reduced number of a number, add the digits.

16 → 7   23 → 5   42 → 6

If you need to, keep adding until you make a 1-digit number.

49 → 13 → 4   28 → 10 → 1

Complete the table of reduced numbers for the multiplication table.

2						5	
							6
			7				

Write about any patterns you notice.

Squares and square roots

Skills summary

- To recognise the square numbers up to 100
- To say what number has been multiplied by itself to produce a given square number, i.e. find its square root
- To square a number which is a multiple of 10, e.g. 40
- To square a 2-digit number
- To find a square root using a calculator
- To find the square of a 1-place decimal number
- To calculate the cubes of numbers up to 10
- To recognise the cube root of the first few cube numbers

Core links

Number Textbook 2 page 62, bottom

- Find, by multiplication, the first 20 square numbers. Use a calculator to check.
- Investigate the differences between consecutive square numbers.

Number Textbook 2 page 63, middle

- Write some 1-place decimal numbers. Find the squares of each by multiplying. Use a calculator to check your answers.

Number Textbook 2 page 64, top

- Write each area in both cm^2 and m^2.

Photocopy Master 87

- Extend the table to include your own squares and square roots, e.g. other multiples of 10, the multiples of 5, ...

Photocopy Master 88

- Extend the table to include the squares of 105, 115, etc. Do similar patterns occur?

Photocopy Master 89

- Write your own 3-digit and 4-digit numbers. Estimate the square roots of the numbers. Use a calculator to find the accurate result, and round it to the nearest whole number. Score the difference between this and your estimate.

Challenge Master 24

Extension Activities

- Investigate 'sneaky' numbers. Use a similar procedure as for finding happy numbers, but find the difference between the two squares, rather than the sum. If the sequence of stages ends at zero, then the number is 'sneaky'. Are all numbers 'sneaky'?

Challenge Activities

- Make a set of cards, and write a different square number on each, including the squares of numbers from 1 to 20, and of multiples of 10. Shuffle the cards and spread them out face down. Reveal the cards, one at a time, and try to say the square root of the revealed number. How many can you answer correctly?
- Investigate the units digit of square numbers. Can you find a square number with a units digit of 0, of 1, of 2, ...?

Name ______________________________

N41

Happy numbers

Some numbers are happy numbers. Try this test to find happy numbers.

If the chain ends at  1 then the number is happy.

32

Square each digit, then add.

9 + 4 = 13

Repeat until you get a 1-digit number.

1 + 9 = 10

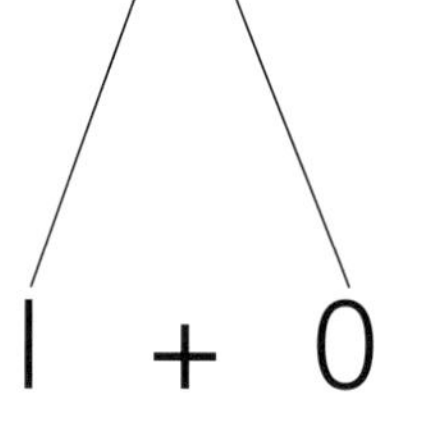

1 + 0 = 1

32 is happy.

How many happy numbers can you find up to 100?

Which numbers are not happy?

Length and perimeter

Skills summary

- To estimate and measure lengths in whole centimetres and millimetres, using a ruler
- To estimate and measure lengths in whole metres
- To draw a line of given length in centimetres and millimetres
- To recognise the relationship between millimetres, centimetres, metres and kilometres
- To convert from one unit of length to another
- To record lengths using decimal notation, e.g. 5·6 cm
- To recognise the relationship between imperial units of length (yard, foot, inch)
- To recognise the relationship between metric and imperial units of length, e.g. kilometres and miles
- To recognise the formula for the perimeter of a rectangle in words and symbols
- To calculate the perimeter of a rectangle, and of shapes drawn on rectangular grids
- To estimate distances in miles and in kilometres
- To understand the concept of speed and its relation to distance and time

Core links

Shape, Data and Measures Textbook page 3, top

- Draw a conversion graph for converting between inches and centimetres. Mark the point (24 inches, 60 cm) to help draw the straight line. Check each calculation on the conversion graph.

Shape, Data and Measures Textbook page 4, top

- If the width of each garden is half its length, find the area of each garden in both square feet and square yards.

Shape, Data and Measures Textbook page 19, top

- Write each perimeter in metres, rounded to the nearest tenth of a metre.

Shape, Data and Measures Textbook page 20, top

- Find the percentage of each page which has been cut out.

Photocopy Master 94

- Write the difference between each pair.

Challenge Master 25

Extension Activities

- Draw a bar-line graph to show the distance of each place from London. Draw horizontal lines to show the limits of 200 km and 500 km.
- Draw a conversion graph to show the relationship between miles and kilometres, marking each of these points on the graph (50 miles is approximately 80 kilometres).
- Use a motoring guide to find similar information and analyse the road distances to and from other towns and cities.

Name ______________________________

M1 M6

Distances from London

Find out the missing distances in miles of the towns and cities from London.

Convert each distance into kilometres.

Write which of these are between 200 km and 500 km from London.

	Miles	Kilometres
Aberdeen	502	
Birmingham		
Bristol	115	
Cardiff		
Carlisle	301	
Dover		
Edinburgh	378	
Exeter		
Glasgow	397	
Holyhead		
Inverness	536	
Leeds		
Liverpool	202	
Manchester		
Newcastle	274	
Norwich		
Nottingham	122	
Plymouth		
Sheffield		

Investigate other large towns and cities that are between 200 km and 500 km from London. Use a map to help you.

Weight and capacity

Skills summary

- To weigh objects in units of 100 grams, and in units smaller than 100 g, e.g. 10 g
- To weigh objects in kilograms and grams
- To recognise the relationship between metric units of weight, (grams and kilograms, kilograms and tonnes)
- To convert gram weights to kilograms and grams, and kilogram and gram weights to grams
- To estimate weights using grams and kilograms
- To relate grams to ounces, kilograms to pounds
- To understand the concept of capacity as a measure of the amount a container will hold
- To measure the capacity of a container in litres, in units of 100 millilitres, in units of 10 centilitres
- To recognise the relationship between metric units of capacity, (litres, millilitres and centilitres)
- To convert from one metric unit of capacity to another
- To relate metric units to imperial units (litres and pints, litres and gallons)

Core links

Shape, Data and Measures Textbook page 6, top

- Calculate the total weight of all the puppies in kilograms. Convert this to pounds.

Shape, Data and Measures Textbook page 7, top

- Draw a conversion graph for converting between pounds and kilograms. Mark the point (22 lb, 10 kg) to help draw the straight line. Check each calculation on the conversion graph.

Shape, Data and Measures Textbook page 9, top

- Find the total capacity of all 11 containers in litres, then express it in gallons and pints.

Photocopy Master 95

- Write the difference between each pair.

Photocopy Master 96

- Write the difference between each pair.

Challenge Master 26

Extension Activities

- Is it possible to find one entry for each letter of the alphabet? How many out of 26 can be found?
- Repeat the activity using a different aspect of mathematics, e.g. shapes. Write the names of different shapes beginning with each of the letters of the alphabet.

Challenge Activity

- Collect labels from food containers, bottles, tins etc. Display these and label them using different units of weight and capacity

Name ___________________________________

Measures alphabet

How many units of measurement do you know?
Write down as many as you can in the table below.
Each unit must begin with the letter on the left.

	Units of measurement
A	
B	
C	
D	decimetre
E	
F	
G	
H	
I	
J	
K	
L	
M	
N	
O	
P	pint
Q	
R	
S	
T	
U	
V	
W	
X	
Y	
Z	

For each unit of measurement, say how it relates to another unit.

Area

Skills summary

- To measure the area of a rectangle by counting squares
- To recognise the formula for the area of a rectangle in words and symbols
- To calculate the area of a rectangle
- To estimate and measure the area of an irregular shape by counting squares
- To use square centimetres, square metres as units of area, and record as m^2, cm^2
- To recognise the relationship between m^2 and cm^2
- To calculate the areas of compound shapes that can be split into rectangles
- To find the area of a right-angled triangle by considering it as half a rectangle
- To calculate the area of a non-right-angled triangle by relating it to its surrounding rectangle
- To calculate the area of a shape by dissecting it into rectangles and triangles
- To calculate the area of a trapezium, a parallelogram
- To calculate the approximate area of a circle

Core links

Shape, Data and Measures Textbook page 12, top and bottom

- Calculate the perimeter of each shape.

Shape, Data and Measures Textbook page 13, top

- Write each area in square metres.
- Calculate the perimeter of each shape.

Shape, Data and Measures Textbook page 14, top

- Find the volume of each cuboid.

Shape, Data and Measures Textbook page 15, top

- Write each area in a different unit, for example, if it is calculated in cm^2, convert it into mm^2 or m^2.

Shape, Data and Measures Textbook page 17, Explore

- Use a similar technique to relate the area of a parallelogram to the area of a rectangle with the same base.

Photocopy Master 97

- Investigate which numbers appear most often on the grid. Which numbers up to 50 do not appear at all?

Photocopy Master 98

- Find the volume of each cuboid.

Challenge Master 27

Extension Activity

- Try the same activity with shapes drawn on isometric spotty paper.

Challenge Activity

- Draw a set of non-right-angled triangles. Draw the 'heights' using a set square, dividing the triangles into two right-angled triangles. Measure the sides of each right-angled triangle, and calculate their areas. Use a calculator if necessary.

Name ______________________________

Dots and areas

For each shape, write:

- **A** its area
- **I** the number of inside dots
- **P** the number of dots on its perimeter.

The space between dots is 1 cm.
Draw some of your own shapes and continue the table.

Shape	A	I	P
1	12 cm^2	6	14
2			
3			

Do you notice a relationship between A, I and P?

Angle

Skills summary

- To recognise and describe angles as acute, obtuse or reflex
- To measure acute angles and obtuse angles using a protractor
- To draw angles using a protractor
- To estimate angles in degrees
- To relate angles (in degrees and right angles) to the 8-point compass and the turning of clock hands
- To calcuate angles in a straight line, given another angle(s)
- To calculate angles at a point, given another angle(s)
- To recognise the sum of the angles of a triangle
- To calculate an angle of a triangle, given the other two
- To recognise the angle sum of a polygon
- To draw a triangle given the three angles

Core links

Shape, Data and Measures Textbook page 22

- Calculate the percentage error of each estimate, i.e. express the error as a percentage of the true angle.

Shape, Data and Measures Textbook page 25, top

- Measure the perimeter of each triangle.

Shape, Data and Measures Textbook page 25, bottom

- Extend to drawing five quadrilaterals, measuring their angles, and finding the total for each.
- Extend the above to pentagons and hexagons.

Photocopy Master 103

- Draw some large triangles. Measure the size of two of the angles. Calculate the size of the third angle. Check by measuring it.

Photocopy Master 104

- Write a formula for the angle total (in terms of the number of right angles) based on n where n = the number of sides.

Challenge Master 28

Extension Activity

- Starting at 12 o'clock, investigate, as accurately as you can, the time when the acute angle between the hands of the clock is 10°, 20°, 30°, 40°, ...

Challenge Activities

- Investigate how many different triangles you can create if all three angles are multiples of 10°.
- If a polygon has n sides, write the angle total in terms of number of right angles. Explore the sizes of the angles of regular polygons.

Name ________________________________

S1 S2

Clock angles

o'clock

Use a clock face with movable hands.

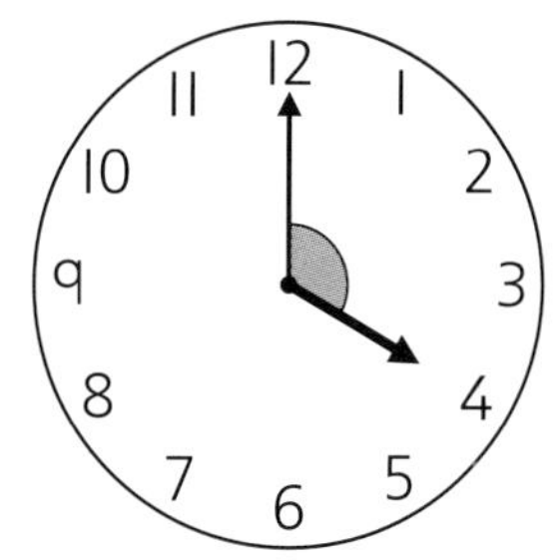

Set the clock to show 4 o'clock.

What is the smaller angle between the hands of the clock?

 Find the angle between the hands for each of the o'clock times.

half past

Set the clock to show 1:30.

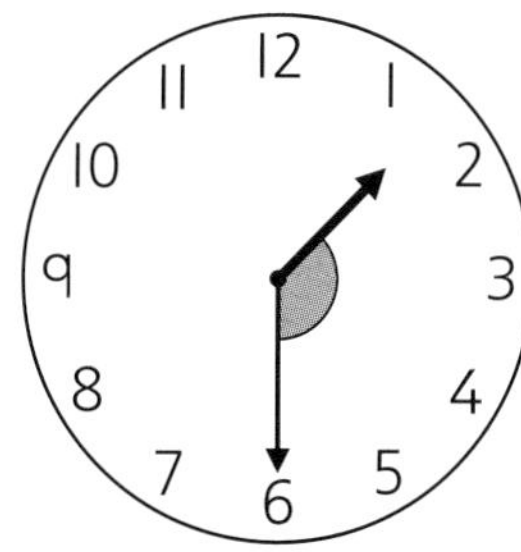

Note that the minute hand points at 6.
The hour hand points halfway between 1 and 2.

What is the smaller angle between the hands of the clock?

 Find the smaller angle between the hands for the other half past times.

quarter past / quarter to

 Investigate the smaller angle between the hands at quarter past and quarter to times.

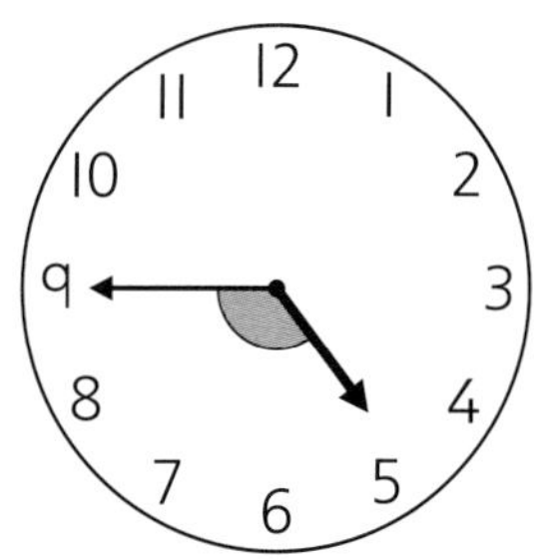

 Investigate for other times.

Coordinates

Skills summary

- To state the coordinates of a point on a coordinate grid
- To plot points on a coordinate grid given their coordinates
- To construct a shape given the coordinates of its vertices
- To define a shape by stating the coordinates of its vertices
- To use and understand the terms: vertical axis, horizontal axis, vertical coordinate, horizontal coordinate
- To recognise the term x-axis and y-axis, and the term 'origin' for the point (0, 0)
- To recognise and plot of points on a coordinate grid with four quadrants
- To reflect a shape in an axis on a coordinate grid, and to recognise the effect on its coordinates.
- To rotate a shape about (0,0) on a coordinate grid, and to recognise the effect on its coordinates.
- To construct straight line graphs of algebraic functions on a coordinate grid, e.g. $y = x + 3$

Core links

Shape, Data and Measures Textbook page 28

- Investigate the objects on the grid which have a horizontal coordinate greater than their vertical coordinate. What do you notice about them?

Shape, Data and Measures Textbook page 29, top

- Investigate the coordinates of these shapes when they are reflected in the horizontal axis, and when reflected in the vertical axis.

Shape, Data and Measures Textbook page 29, bottom

- Find the area of each shape.

Shape, Data and Measures Textbook page 30, Explore

Investigate points whose coordinates differ by 2, differ by 3, ...

Photocopy Master 105

- Vary the game so that, for example, the winner is the first to complete the four corners of a square.

Challenge Master 29

Extension Activity

- Investigate the effects on the coordinates of rotating the shape through turns of 90° about the origin, (0,0).

Name ______________________________

Reversing signs

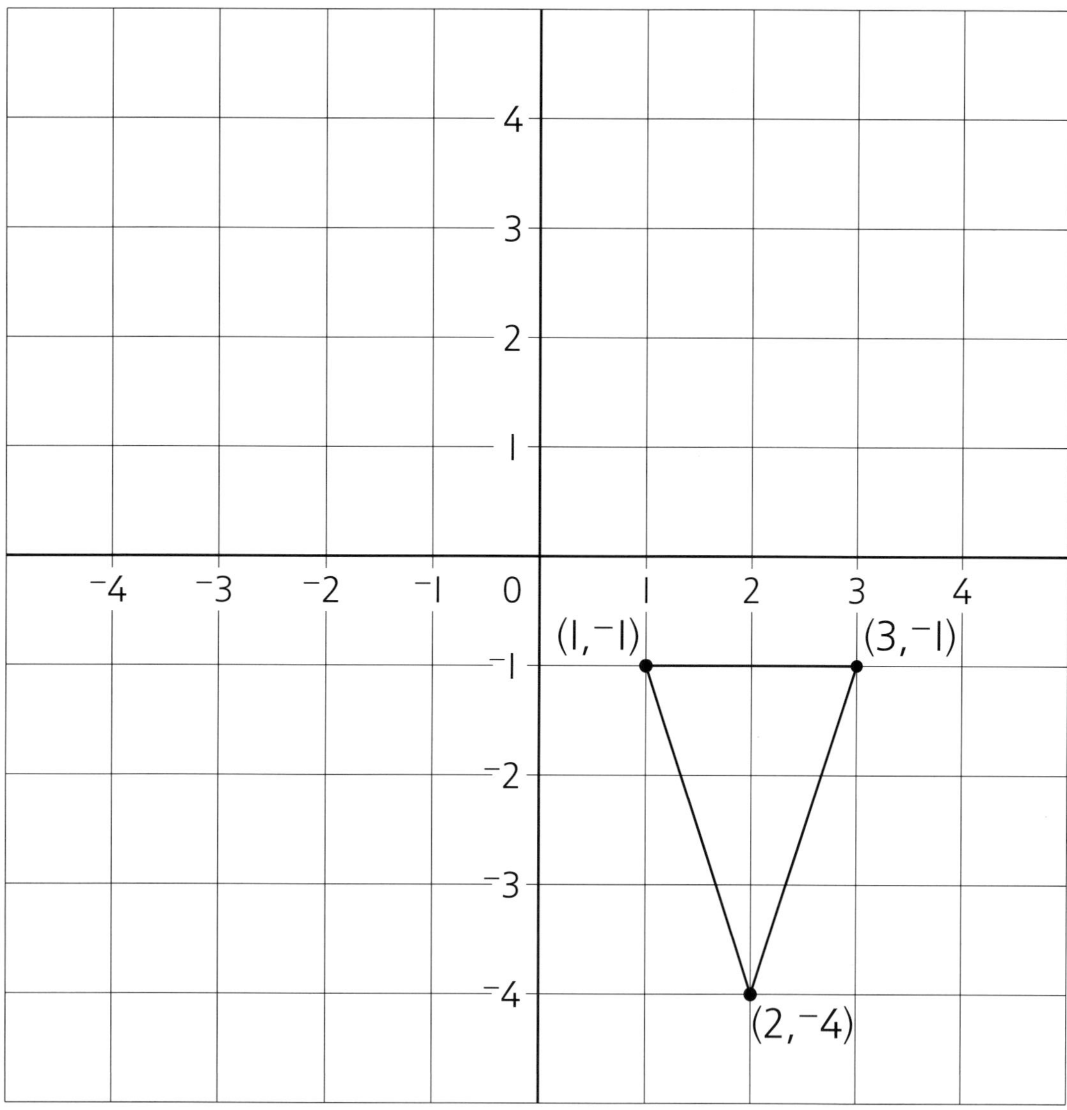

The coordinates of the triangle are $(1,{}^{-}1)$ $(3,{}^{-}1)$ $(2,{}^{-}4)$.

Investigate reversing the signs of:

a the vertical coordinates
b the horizontal coordinates
c both coordinates.

Draw the new triangle each time.

 Investigate the same activity but starting with a different shape in a different position.

Reflection, rotation and translation

Skills summary

- To locate a line of symmetry on a pattern or shape, including regular polygons
- To draw shapes and patterns with line symmetry
- To complete a symmetrical pattern/drawing, given a line of symmetry and half the pattern/drawing
- To draw the reflection of a shape in a mirror line parallel to one edge
- To draw the reflection of a shape in a mirror line when the sides of the shape are not parallel or perpendicular to one edge
- To complete symmetrical patterns based on two lines of symmetry at right angles
- To recognise rotations of shapes about a vertex after quarter- and half-turns, and 45° turns
- To sketch the position of polygons on a coordinate grid after rotations of 180° and 90°
- To sketch the position of a shape after a translation, and to define a translation
- To understand the meaning of rotational symmetry
- To construct shapes and patterns with rotational symmetry
- To recognise the order of rotational symmetry of shapes

Core links

Shape, Data and Measures Textbook page 32, bottom

- Draw the reflection of each shape in both axes.

Shape, Data and Measures Textbook page 34, top

- Describe the translation for each point to its new position when it has been reflected in the x-axis, in the y-axis, and when it has been rotated 90° and 180° about the origin.

Shape, Data and Measures Textbook page 34, bottom

- Describe two reflections that are equivalent to each translation.

Photocopy Master 107

- Write a rule based on the coordinates of a point, when it is reflected in the x-axis, the y-axis and in both axes.

Challenge Master 30

Extension Activities

- Extend the activity to 6 × 6 or 8 × 8 square grids.
- Make patterns with rotational symmetry on a 5 × 5 spotty arrangement.

Name ___________________________

S4 S5

Rotational symmetry patterns

Make patterns with rotational symmetry on 4 × 4 square grids.
Start at the centre and draw either of these pairs of lines.

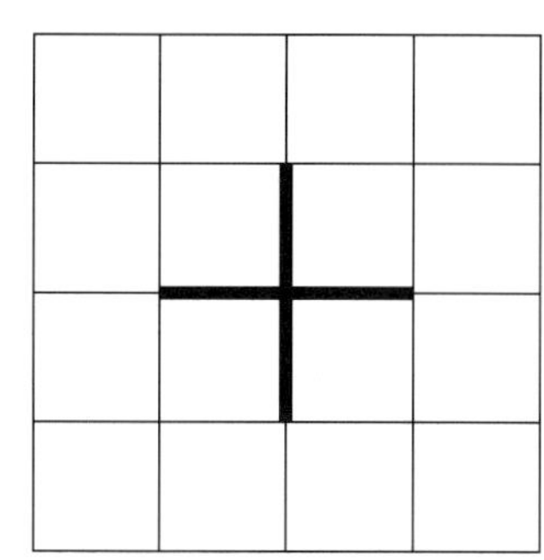
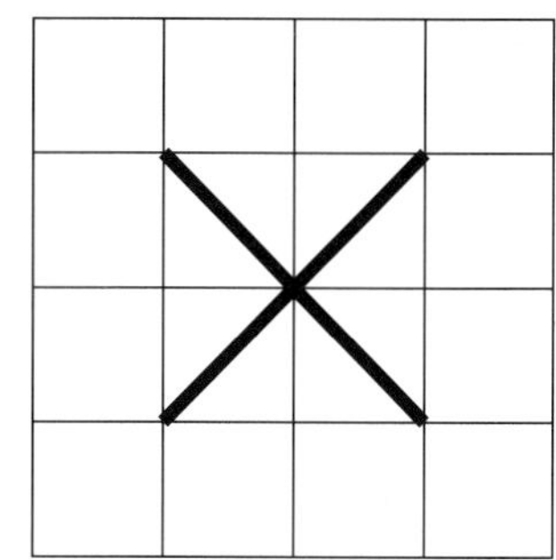

Here are two rotational symmetry patterns.

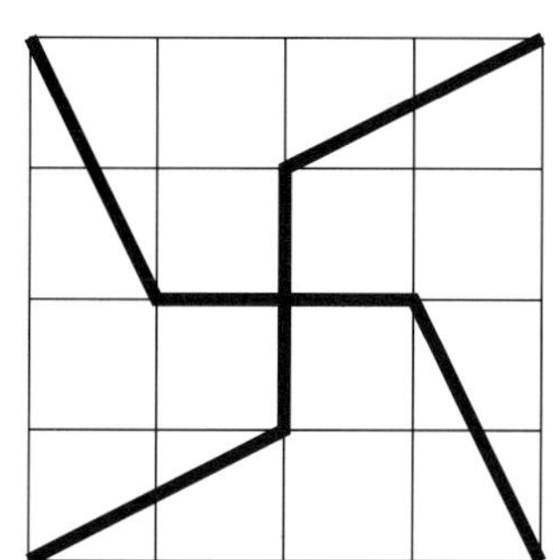
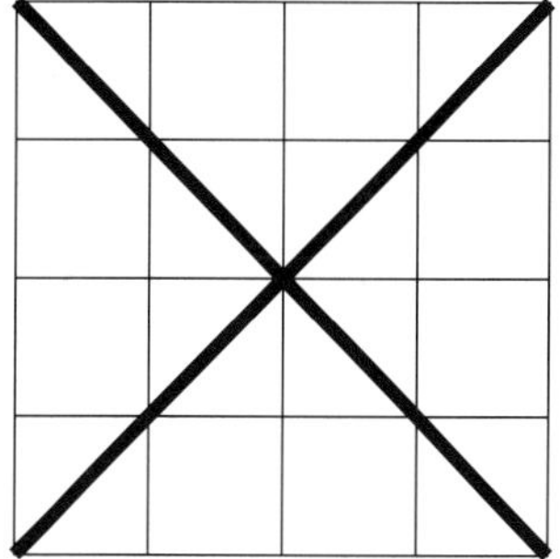

Draw some of your own.

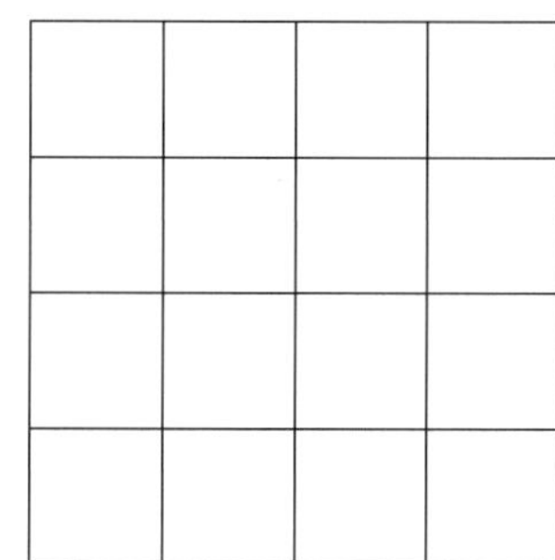
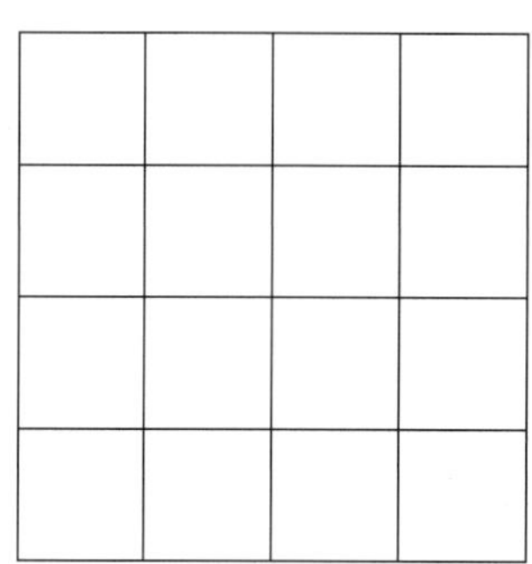

3-d shape

Skills summary

- To recognise the names and properties of 3-d shapes: cube, cuboid, pyramid, sphere, cone, cylinder, prism
- To recognise and describe different types of pyramid: triangle-based, square-based, ...
- To recognise and describe different types of prism: pentagonal, hexagonal, ...
- To construct a net for a given shape
- To recognise a polyhedron as a 3-d shape whose faces are polygons
- To recognise tetrahedra, octahedra and dodecahedra as 4-, 8- and 12-sided polyhedra
- To classify 3-d shapes according to their properties, e.g. faces, vertices, edges
- ↓ To develop the relationship between the faces, edges and vertices of a 3-d shape
- To recognise parallel and perpendicular faces on 3-d shapes
- ↓ To locate planes of symmetry on 3-d shapes
- ↓ To recognise regular polyhedra

Core links

Shape, Data and Measures Textbook page 37, top

- Write how many parallel faces each shape has.
- For each shape, record the number of faces, vertices and edges in a table. Look for a general relationship connecting the three numbers.

Shape, Data and Measures Textbook page 38, Explore

- Investigate how many different nets of cubes can be found.

Shape, Data and Measures Textbook page 39, top

- Write the number of edges, vertices and faces for each polyhedron. Use models of the shapes to help you. Explore the relationship between them.

Photocopy Master 109

- Investigate the symmetrical properties of each net.
- Investigate whether or not each net has the same perimeter.

Photocopy Master 110

- Investigate the relationship between the width of the rectangle and the length of the diagonal of the rectangle.

Challenge Master 31

Extension Activity

- Try the same activity with different-sized cuboids, e.g. $2 \times 3 \times 4$.

Name ______________________________

Painting cubes

A $3 \times 3 \times 3$ cube is made by joining 27 small white cubes.

The outside is painted red, and then the 27 cubes are separated.

How many of these cubes will have:

0 red faces?
1 red face?
2 red faces?
3 red faces?

Repeat the activity for different sized cubes, for example, $4 \times 4 \times 4$.

Can you see any patterns?

2-d shape

Skills summary

- To understand the meaning of 'polygon'
- To name polygons: square, rectangle, triangle, pentagon, hexagon, heptagon, octagon, nonagon, decagon
- To recognise the difference between regular and irregular polygons
- To construct polygons in different ways, e.g. by drawing, by cutting, using a geoboard
- To recognise and draw the diagonals of a polygon
- To recognise angle properties of regular polygons
- To recognise horizontal and vertical lines, parallel and perpendicular lines
- To recognise parallel and perpendicular sides in a 2-d shape
- To draw a line parallel or perpendicular to a given line
- To recognise the properties of parallelogram, a rhombus, a trapezium
- To recognise and classify different types of quadrilateral

Core links

Shape, Data and Measures Textbook page 40, top

- Label each side of each shape with a letter. Write pairs of perpendicular sides using the letters.

Shape, Data and Measures Textbook page 41, top

- Write the area of each shape.

Shape, Data and Measures Textbook page 41, Explore

- Draw a different shaped rectangle, and mark the point in a different position.

Shape, Data and Measures Textbook page 42, top

- Write the area of each shape.

Shape, Data and Measures Textbook page 48, top

- Create a different 3-piece dissection of a square, cut out the pieces, and investigate different shapes that can be created by joining the pieces along equal sides.

Photocopy Master 111

- Find the area of each of your created shapes.

Challenge Master 32

Extension Activity

- Devise a set of anagrams for words which relate to the properties of shapes. Possible words include: horizontal, vertical, parallel, perpendicular, radius, diameter, circumference, side, face, edge, vertex, straight, curved, regular, polygon, angle, right angle, protractor, ...

Name ______________________________

S7 **S9**

Shape anagrams

Rearrange the letters to make the name of a shape.

1 quaser ______________

2 buce ______________

3 tracengle ______________

4 oxhagen ______________

5 drincley ______________

6 shurbom ______________

7 repesh ______________

8 lapmelgrolara ______________

9 glantier ______________

10 once ______________

11 buodic ______________

12 tagponen ______________

13 contago ______________

14 smirp ______________

15 lericc ______________

16 draimpy ______________

Invent your own set of mathematical anagrams.

Quadrilaterals

Skills summary

- To recognise isosceles triangles as having two equal sides, and two equal angles
- To recognise equilateral triangles as having three equal sides and three equal angles
- To recognise right-angled triangles, and scalene triangles
- To classify triangles according to their types and properties
- To classify triangles into acute-angled, obtuse-angled and right-angled
- To recognise the properties of square, rectangle, parallelogram, rhombus, trapezium
- To recognise the properties of a kite
- To recognise the properties of an arrowhead
- To classify quadrilaterals according to their types and properties

Core links

Shape, Data and Measures Textbook page 43, Explore

- Using isometric dotty paper, draw different rhombuses, and repeat the activity of drawing and exploring the properties of the diagonals.

Shape, Data and Measures Textbook page 44, top

- Write the area of each shape (consider the space between pegs to be 1 cm).

Shape, Data and Measures Textbook page 44, Explore

- It is possible to create 25 different pentagons. How many can you find?

Photocopy Master 113

- Write the names of the unshaded shapes.

Challenge Master 33

Extension Activities

- Investigate the number of different kites, with whole number diagonal lengths, which can be drawn with areas 36 cm^2, 20 cm^2, and so on.
- Investigate a similar relationship for arrowheads.

Name ______________________________

Kites

Use spotty paper.

Draw several different kites.

Draw both diagonals for each.

Write the lengths of the diagonals in the table below.

Find the areas of the kites by splitting them up into part squares and part rectangles.

Complete the table below.

Can you find a relationship between the area and the lengths of the diagonals?

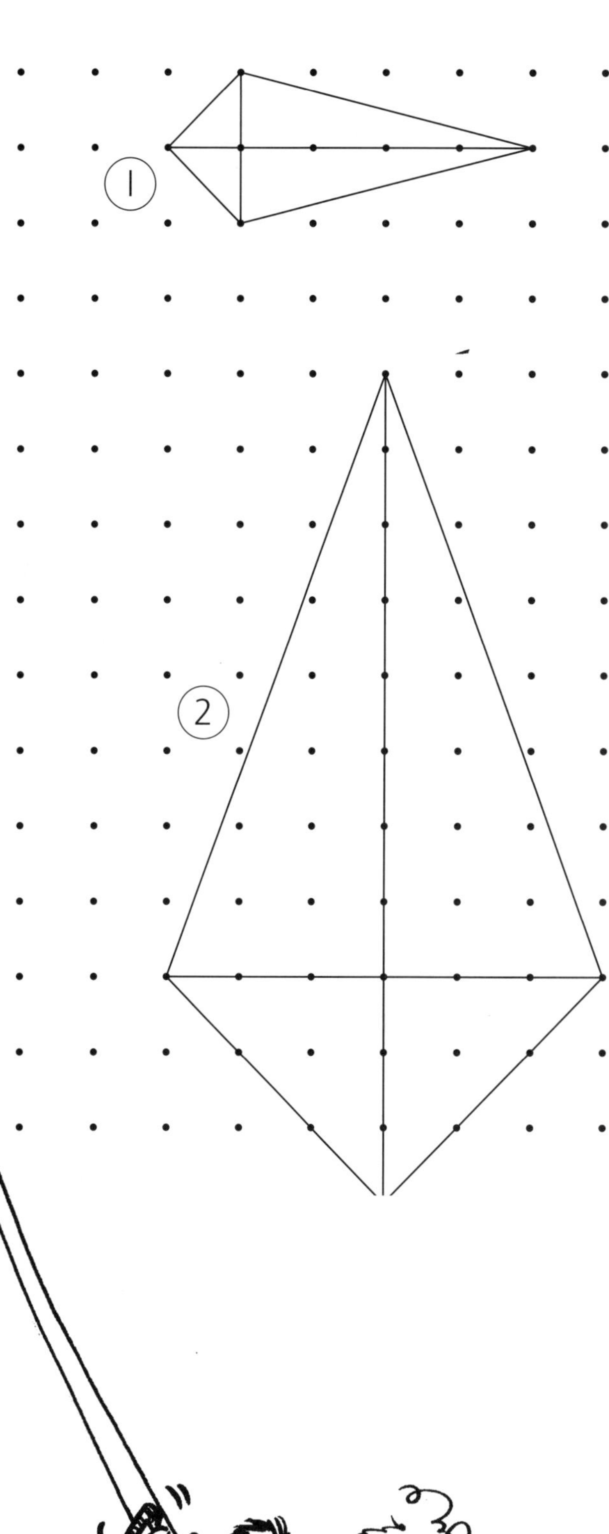

Kite	Diagonals	Area
1	2 cm, 5 cm	5 cm^2
2		
3		

D1 Grouped data **D2** Pie charts **D3** Conversion graphs

Graphs

Skills summary

- To interpret a bar-line graph
- To draw a bar-line graph
- To interpret a line graph, e.g. temperature graph
- To construct and interpret a pie chart
- To use a line graph to approximate the value of one variable, given the other
- To draw a line graph
- To construct and interpret a conversion graph
- To group discrete data in equal intervals
- To construct a frequency table based on grouped discrete data
- To draw and interpret a bar graph based on grouped discrete data
- To group continuous data in equal intervals
- To construct a frequency table based on grouped continuous data
- To draw and interpret a histogram

Core links

Shape, Data and Measures Textbook page 49

- Choose a different way of grouping the data, e.g. 1–4, 5–8, 9–12. Draw a bar graph, and compare it with the previous graph.

Shape, Data and Measures Textbook page 50, top

- Choose a different way of grouping the data. Draw a bar graph, and compare it with the previous graph.

Shape, Data and Measures Textbook page 52, top and bottom

- Write the fraction of the total votes for each television programme and film.
- Write the percentage of the total votes for each television programme and each film.

Shape, Data and Measures Textbook page 52, bottom

- Collect similar data from 16 children and show the results on a pie chart.

Shape, Data and Measures Textbook page 53, top

- Write the fraction of the total votes for each sport.

Shape, Data and Measures Textbook page 53, bottom

- Repeat the activity but roll the dice 32 times, and draw a pie chart to show the results on a circle with 16 equally-spaced divisions.

Shape, Data and Measures Textbook page 55, bottom

- Collect data on current exchange rates between pounds and another currency. Draw a conversion graph and use it to record some conversions.

Shape, Data and Measures Textbook page 56, bottom

- Collect data from a newspaper showing current temperatures in different parts of the world. Convert these from one unit of temperature to another.

Challenge Master 34

Extension Activity

- Superimpose a different journey on the same graph, and write a story for it.

Name ______________________________

Cycle ride

Our cycle ride

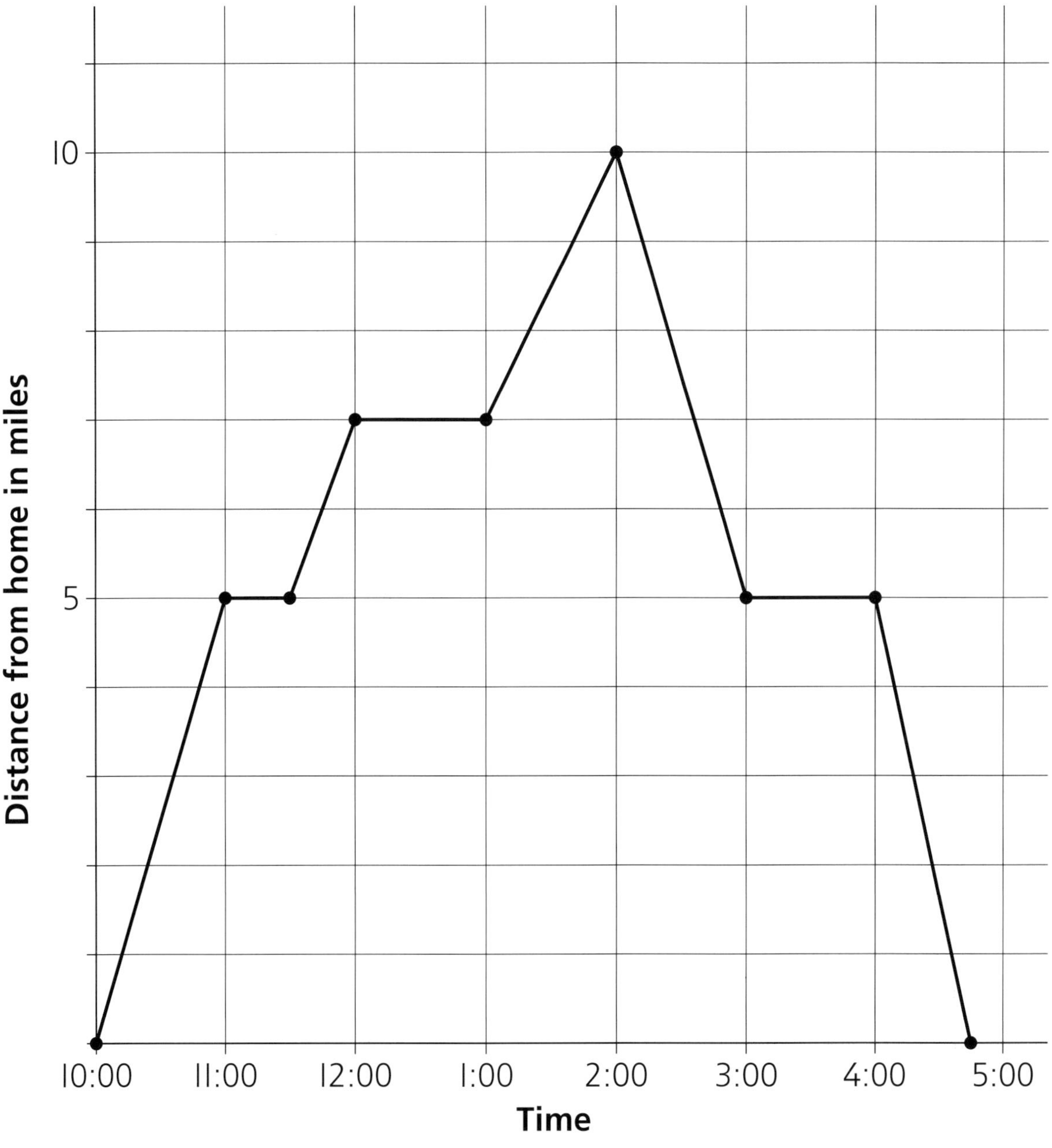

Distance from home in miles

10

5

10:00 11:00 12:00 1:00 2:00 3:00 4:00 5:00

Time

Invent a story to match this graph.

Write about the places visited at each stage of the journey, and the time taken.

Use graph paper to draw another journey of your own.

Averages

Skills summary

- To understand the concept of average as a 'middle' number
- To understand the 'mode' of a distribution as the observation which occurs most frequently
- To recognise the median as the 'middle' observation, when they are placed in ascending order
- To find the median of a set of discrete data (odd number of data items)
- To find the median of a set of discrete data (even number of data items)
- To recognise and calculate the mean of a set of observations
- To compare quantities as being above or below the mean
- To recognise the mean, median and mode as three types of average
- To estimate the mean of a set of data
- To find the mean of a frequency distribution based on discrete data
- To find the mean of a frequency distribution based on grouped discrete data
- To find the mean of a frequency distribution based on grouped continuous data

Core links

Shape, Data and Measures Textbook page 59, middle

- Repeat the activity, but pick one more 1-digit number to add to each child's set of scores so that they have each had 4 turns. Convert the means which have remainders into decimal form.

Shape, Data and Measures Textbook page 60, top

- Collect data on shoe sizes for pupils in the class/school. Ideally, use a foot measurer like those used by shoe shops. Calculate the modal shoe size.

Shape, Data and Measures Textbook page 61, Explore

- Repeat the activity, but find the difference between the two dice throws instead of the total.

Photocopy Master 121

- Extend to four cards and five cards.
- Repeat the activity, but choosing a different set of cards to include odd numbers, e.g. three of each of 2, 3, 4, 5 and 6.

Challenge Master 35

Extension Activity

- Collect data from a newspaper relating to house prices and calculate mean prices.

Challenge Activity

- Collect data from a newspaper showing football scores. Use these to calculate the mean number of goals scored per team, together with the mode and median. How do they compare?

Name ______________________________

D4

House prices

Seaville

				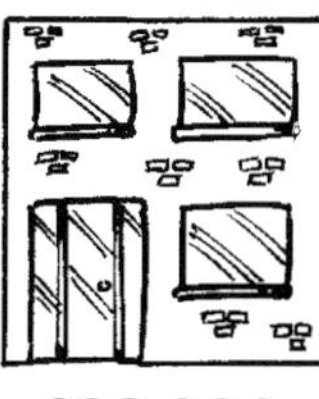
 £135 000	£116 500	£120 000	£101 500	£83 000

£77 500	£103 000	£111 500	£125 000	£98 000

				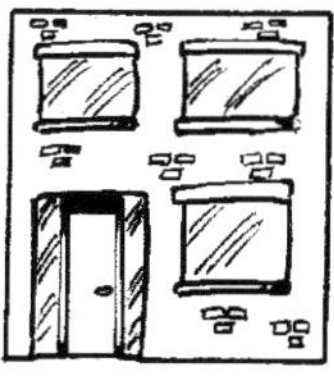
£108 500	£93 000	£79 000	£84 500	£75 000
				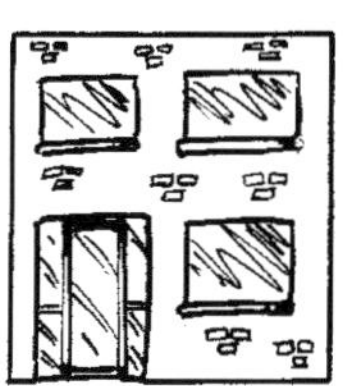
£79 500	£88 500	£85 000	£77 500	£92 000

Estimate, then calculate, the mean house price, first in Seaville, then in Oldtown.
Use a calculator to check your answers.
How do they compare?

Collect data about house prices from a local newspaper, and find the mean house price.

Probability

Skills summary

- To classify events as impossible, unlikely, likely, certain
- To classify events as impossible, very unlikely, unlikely, likely, very likely, certain
- To recognise equally likely events
- To predict the outcome of equally likely events based on proportional frequencies
- To define the chances associated with equally likely events
- To define the chances associated with non-equally likely events
- To recognise the probability scale from 0 to 1
- To assign probabilites to equally likely events
- To assign probabilites to non-equally likely events
- To recognise that if the probability of an event occurring is p, then the probability of it not occurring is 1 – p
- To recognise non-equally likely events
- To predict the outcome of non-equally likely events based on proportional frequencies
- To list all the possible outcomes of an event

Core links

Shape, Data and Measures Textbook page 62, top

- Throw a dice 36 times and show how many throws were in the following four categories:
  (a) 1
  (b) 2
  (c) 3 or 4
  (d) 5 or 6.
  Start by predicting how many you think will be in each category.

Shape, Data and Measures Textbook page 63, top

- Consider a 10-sided dice numbered from 1 to 10. Write some events and associated probabilities.

Shape, Data and Measures Textbook page 64, bottom

- Use a different number of coloured cubes placed in a bag, and decide on a number of draws from the bag. Repeat the activity of predicting how many of each colour will be drawn, then test your predictions. Write the probability of drawing each colour.

Photocopy Master 123

- Devise similar games based on throwing a 10-sided dice, or spinning a 1–10 spinner. Make and write down predictions at the beginning of each game.

Challenge Master 36

Extension Activities

- Draw a new racetrack based on three 'horses': 'difference is even', 'difference is odd', and 'difference is 0'. Play the game and investigate the probabilities of each 'horse' winning.
- Extend to using dice numbered differently, and exploring the new differences and their probabilities.

Name ______________________________

Difference race

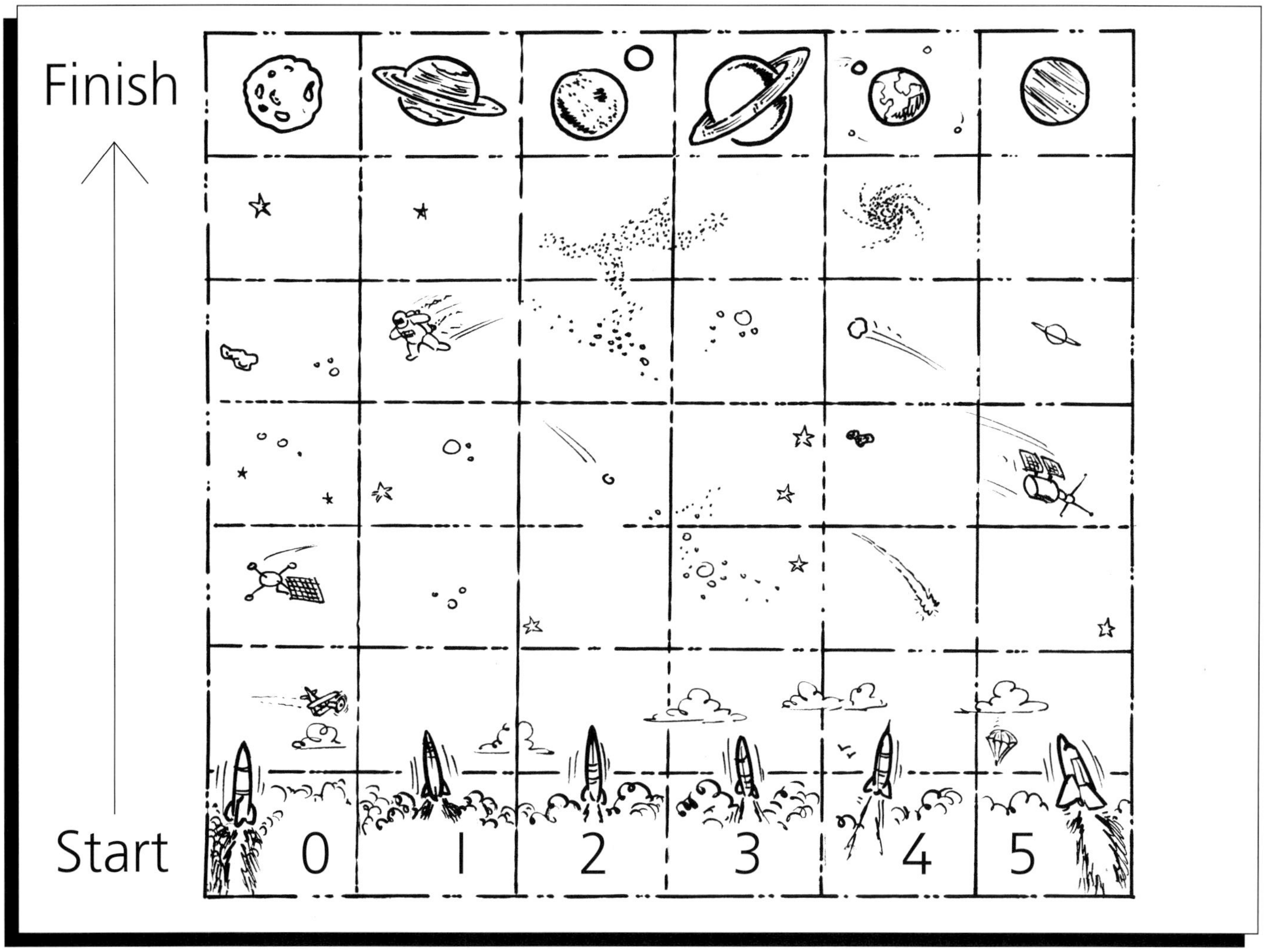

Finish

Start

0	1	2	3	4	5

A game for two players.

You need two dice (one red and one blue), and six counters.
Place a counter at the start of each track.

Take turns to throw the two dice and find the difference between the scores.

Move the counter whose number matches the difference, one space forward.

Continue until one counter reaches 'Finish'.
The person who moves last wins. Play several games.

Draw a difference table to show all the possible differences.

Write the probability for each difference.

Red (columns), **Blue** (rows)

d	1	2	3	4	5	6
1						
2			1			
3						
4						
5						
6	5					

Challenge Master Answers

Challenge Master 1

1.	4·63	**2.**	9·46
3.	3·64	**4.**	6·49
5.	3·96	**6.**	3·69
7.	4·93	**8.**	6·93 or 6·94
9.	3·46 or 3·49	**10.**	4·69
11.	6·34	**12.**	4·36 or 4·39
13.	9·36 or 9·43	**14.**	9·34
15.	3·94	**16.**	4·96

Challenge Master 2

1.	16·2	**2.**	70
3.	3120	**4.**	544
5.	345·6	**6.**	1180
7.	152	**8.**	872
9.	0·391	**10.**	1·465
11.	23·3	**12.**	1·165
13.	0·982	**14.**	0·2909
15.	0·795	**16.**	1·164

Challenge Master 3

2-way splits
$1 \times 9 = 9$
$2 \times 8 = 16$
$3 \times 7 = 21$
$4 \times 6 = 24$
$5 \times 5 = 25$

3-way splits
$1 \times 1 \times 8$
$1 \times 2 \times 7 = 14$
$1 \times 3 \times 6 = 18$
$1 \times 4 \times 5 = 20$
$2 \times 2 \times 6 = 24$
$2 \times 3 \times 5 = 30$
$2 \times 4 \times 4 = 32$
$3 \times 3 \times 4 = 36$

4-way splits
$1 \times 1 \times 1 \times 7$
$1 \times 1 \times 2 \times 6 = 12$
$1 \times 1 \times 3 \times 5 = 15$
$1 \times 1 \times 4 \times 4 = 16$
$1 \times 2 \times 2 \times 5 = 20$
$1 \times 2 \times 3 \times 4 = 24$
$1 \times 3 \times 3 \times 3 = 27$
$2 \times 2 \times 2 \times 4 = 32$
$2 \times 2 \times 3 \times 3 = 36$

Many 5-way splits, of which largest answer $2 \times 2 \times 2 \times 2 \times 2 = 32$

The largest answer is 36. The largest answer for each split comes from dividing the strip into equal parts, or as nearly equal as possible.

Challenge Master 4

1.	$17 \times 45 = 765$	**2.**	$26 \times 33 = 858$
3.	$18 \times 51 = 918$	**4.**	$35 \times 21 = 735$
5.	$46 \times 53 = 2438$	**6.**	$62 \times 34 = 2108$

Challenge Master 5

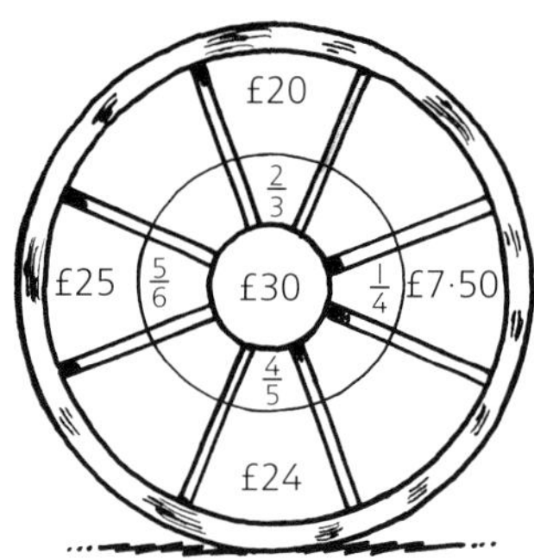

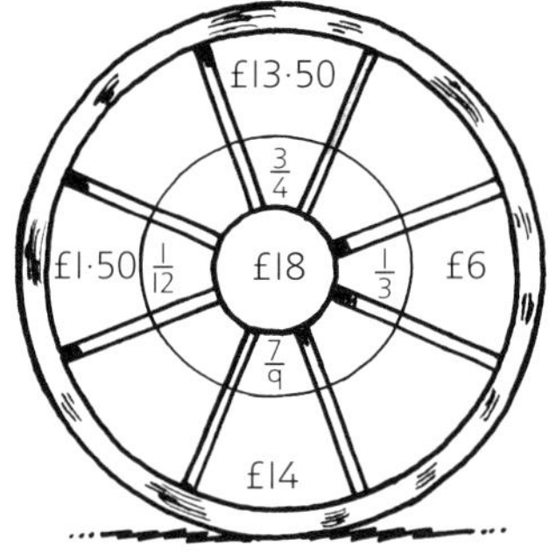

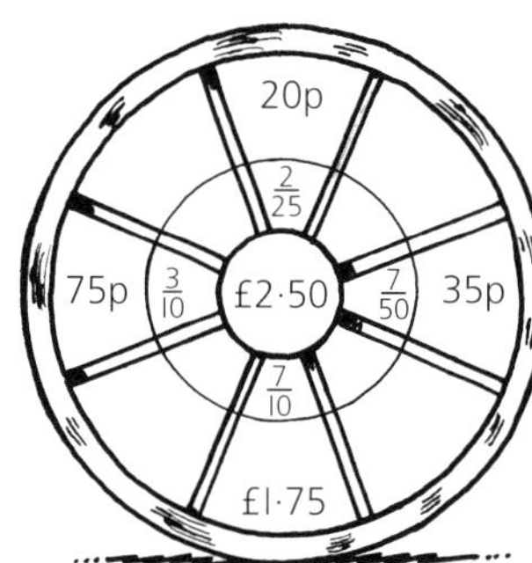

Challenge Master 6

$\frac{1}{2} = \frac{3}{6}$ $\frac{1}{2} = \frac{4}{8}$ $\frac{1}{2} = \frac{5}{10}$ $\frac{1}{2} = \frac{6}{12}$ $\frac{1}{2} = \frac{7}{14}$ $\frac{1}{2} = \frac{8}{16}$ $\frac{1}{2} = \frac{9}{18}$

$\frac{1}{2} = \frac{10}{20}$ $\frac{1}{3} = \frac{2}{6}$ $\frac{1}{3} = \frac{4}{12}$ $\frac{1}{3} = \frac{5}{15}$ $\frac{1}{4} = \frac{2}{8}$ $\frac{1}{4} = \frac{3}{12}$ $\frac{1}{4} = \frac{5}{20}$

$\frac{1}{5} = \frac{2}{10}$ $\frac{1}{5} = \frac{3}{15}$ $\frac{1}{5} = \frac{4}{20}$ $\frac{1}{6} = \frac{2}{12}$ $\frac{1}{6} = \frac{3}{18}$ $\frac{1}{7} = \frac{2}{14}$ $\frac{1}{8} = \frac{2}{16}$

$\frac{1}{9} = \frac{2}{18}$ $\frac{1}{10} = \frac{2}{20}$ $\frac{2}{3} = \frac{4}{6}$ $\frac{2}{3} = \frac{6}{9}$ $\frac{2}{3} = \frac{8}{12}$ $\frac{2}{3} = \frac{10}{15}$ $\frac{2}{3} = \frac{12}{18}$

$\frac{3}{4} = \frac{6}{8}$ $\frac{3}{4} = \frac{9}{12}$ $\frac{3}{4} = \frac{12}{16}$ $\frac{3}{4} = \frac{15}{20}$ $\frac{2}{5} = \frac{4}{10}$ $\frac{2}{5} = \frac{6}{15}$ $\frac{2}{5} = \frac{8}{20}$

$\frac{3}{5} = \frac{6}{10}$ $\frac{3}{5} = \frac{9}{15}$ $\frac{3}{5} = \frac{12}{20}$ $\frac{4}{5} = \frac{8}{10}$ $\frac{4}{5} = \frac{12}{15}$ $\frac{4}{5} = \frac{16}{20}$ $\frac{5}{6} = \frac{10}{12}$

$\frac{5}{6} = \frac{15}{18}$ $\frac{2}{7} = \frac{4}{14}$ $\frac{3}{7} = \frac{6}{14}$ $\frac{4}{7} = \frac{8}{14}$ $\frac{5}{7} = \frac{10}{14}$ $\frac{6}{7} = \frac{12}{14}$ $\frac{3}{8} = \frac{6}{16}$

$\frac{5}{8} = \frac{10}{16}$ $\frac{7}{8} = \frac{14}{16}$ $\frac{2}{9} = \frac{4}{18}$ $\frac{4}{9} = \frac{8}{18}$ $\frac{5}{9} = \frac{10}{18}$ $\frac{7}{9} = \frac{14}{18}$ $\frac{8}{9} = \frac{16}{18}$

$\frac{3}{10} = \frac{6}{20}$ $\frac{7}{10} = \frac{14}{20}$ $\frac{9}{10} = \frac{18}{20}$

Challenge Master 7

1. 10·485 + 0·515 = 11 seconds
2. 11·218 + 0·782 = 12 seconds
3. 9·961 + 0·039 = 10 seconds
4. 12·047 + 0·953 = 13 seconds
5. 11·573 + 0·427 = 12 seconds
6. 13·308 + 0·692 = 14 seconds
7. 9·951 + 0·049 = 10 seconds
8. 10·596 + 0·404 = 11 seconds
9. 10·386 + 0·614 = 11 seconds
10. 11·470 + 0·530 = 12 seconds

Challenge Master 8

9: 4 + 5
2 + 3 + 4

25: 12 + 13
3 + 4 + 5 + 6 + 7

50: 11 + 12 + 13 + 14
8 + 9 + 10 + 11 + 12

36: 11 + 12 + 13
1 + 2 + 3 + 4 + 5 + 6 + 7 + 8

21: 10 + 11
6 + 7 + 8
1 + 2 + 3 + 4 + 5 + 6

15: 7 + 8
4 + 5 + 6
1 + 2 + 3 + 4 + 5

33: 16 + 17
10 + 11 + 12
3 + 4 + 5 + 6 + 7 + 8

60: 19 + 20 + 21
10 + 11 + 12 + 13 + 14
4 + 5 + 6 + 7 + 8 + 9 + 10 + 11

45: 22 + 23
14 + 15 + 16
7 + 8 + 9 + 10 + 11
5 + 6 + 7 + 8 + 9 + 10
1 + 2 + 3 + 4 + 5 + 6 + 7 + 8 + 9

63: 31 + 32
20 + 21 + 22
8 + 9 + 10 + 11 + 12 + 13
6 + 7 + 8 + 9 + 10 + 11 + 12
3 + 4 + 5 + 6 + 7 + 8 + 9 + 10 + 11

75: 37 + 38
24 + 25 + 26
13 + 14 + 15 +16 + 17
10 + 11 + 12 + 13 + 14 + 15
3 + 4 + 5 + 6 + 7 + 8 + 9 + 10 + 11 + 12

Numbers which cannot be made are the powers of 2, i.e. 1, 2, 4, 8, 16, 32, ...

Challenge Master 9

	2	**3**	**4**	**5**	**6**	**7**	**8**	**9**	**10**	**11**
2	2	6	4	10	6	14	8	18	10	22
3	6	3	12	15	6	21	24	9	30	33
4	4	12	4	20	12	28	8	36	20	44
5	10	15	20	5	30	35	40	45	10	55
6	6	6	12	30	6	42	24	18	30	66
7	14	21	28	35	42	7	56	63	70	77
8	8	24	8	40	24	56	8	72	40	88
9	18	9	36	45	18	63	72	9	90	99
10	10	30	20	10	30	70	40	90	10	110
11	22	33	44	55	66	77	88	99	110	11

Challenge Master 10

1. yes **2.** no **3.** no **4.** yes
5. yes **6.** no **7.** yes **8.** yes

For 4-digit numbers, e.g. 3563, double the hundreds, i.e. double 35 = 70, add this to the tens and units, i.e. 70 + 63 = 133.

Challenge Master 11

1.
$9 \times 6 = \underline{54}$
$9 \times 66 = \underline{594}$
$9 \times 666 = \underline{5994}$
$9 \times 6666 = \underline{59994}$

2.
$6 \times 9 = \underline{54}$
$6 \times 99 = \underline{594}$
$6 \times 999 = \underline{5994}$
$6 \times 9999 = \underline{59994}$

3.
$9 \times 9 = \underline{81}$
$99 \times 99 = \underline{9801}$
$999 \times 999 = \underline{998001}$
$9999 \times 9999 = \underline{99980001}$

4.
$6 \times 9 = \underline{54}$
$66 \times 99 = \underline{6534}$
$666 \times 999 = \underline{665334}$
$6666 \times 9999 = \underline{66653334}$

5.
$12 \times 99 = \underline{1188}$
$23 \times 99 = \underline{2277}$
$34 \times 99 = \underline{3366}$
$45 \times 99 = \underline{4455}$

6.
$9 \times 222222 = \underline{1999998}$
$9 \times 333333 = \underline{2999997}$
$9 \times 444444 = \underline{3999996}$
$9 \times 555555 = \underline{4999995}$

7.
$(9 \times 1) + 2 = \underline{11}$
$(9 \times 12) + 3 = \underline{111}$
$(9 \times 123) + 4 = \underline{1111}$
$(9 \times 1234) + 5 = \underline{11111}$

8.
$(9 \times 9) + 7 = \underline{88}$
$(9 \times 98) + 6 = \underline{888}$
$(9 \times 987) + 5 = \underline{8888}$
$(9 \times 9876) + 4 = \underline{88888}$

Challenge Master 12

If the dice numbers are a, b and c then:

1. double the first number — 2a
2. add 2 — 2a + 2
3. multiply by 5 — 10a + 10
4. add the second number — 10a + 10 + b
5. multiply by 10 — 100a + 100 + 10b
6. add the third number — 100a + 100 + 10b + c
7. subtract 100 — 100a + 10b + c

which gives the 3-digit number a b c.

Challenge Master 13

1.	$\frac{3}{5}$	=	0·6	**2.**	$\frac{6}{5}$	=	1·2
3.	$\frac{4}{10}$	=	0·4	**4.**	$\frac{7}{5}$	=	1·4
5.	$\frac{23}{5}$	=	4·6	**6.**	$\frac{12}{10}$	=	1·2
7.	$\frac{9}{5}$	=	1·8	**8.**	$\frac{8}{5}$	=	1·6
9.	$\frac{17}{2}$	=	8·5	**10.**	$\frac{44}{8}$	=	5·5

Challenge Master 14

Fractions with a denominator of:

2 and 5	can produce 1-place decimals
4	can produce up to 2-place decimals
8	can produce up to 3-place decimals
3, 6, 7 and 9	can produce recurring decimals

Challenge Master 15

Answers are as follows:

3. f + e = 2·784
9. a + h + f = 10·494
12. e + f + c + a = 23·734
14. h + c + e + b = 25·932
15. a + b + f + h = 12·194
16. c + e + f + b = 20·484
17. a + d + e + h = 48·982

Challenge Master 16

1.
0·1
2·4 | 2·5
3·8 | 1·4 | 3·9
5·3 | 1·5 | 2·9 | 6·8

2.
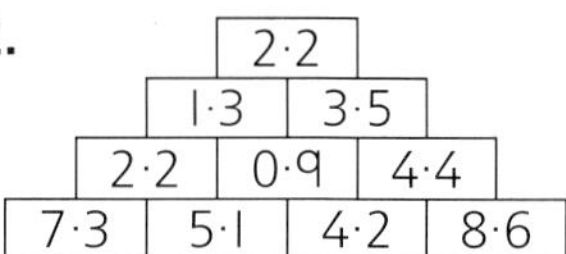
2·2
1·3 | 3·5
2·2 | 0·9 | 4·4
7·3 | 5·1 | 4·2 | 8·6

3.
1·01
0·49 | 1·50
3·85 | 3·36 | 1·86
5·21 | 1·36 | 4·72 | 6·58

4.
2·27
3·69 | 1·42
7·58 | 3·89 | 5·31
7·96 | 0·38 | 4·27 | 9·58

5.
0·66
2·8 | 2·14
1·82 | 4·62 | 2·48
4·5 | 6·32 | 1·7 | 4·18

6.
0·95
1·27 | 2·22
1·59 | 2·86 | 0·64
0·31 | 1·9 | 4·76 | 5·4

Challenge Master 17

Possible answers include:

2 is a factor of 24	2 is a factor of 14
3 is a factor of 45	3 is a factor of 33
4 is a factor of 52	4 is a factor of 16
5 is a factor of 25	5 is a factor of __
6 is a factor of 48	6 is a factor of 36
7 is a factor of 28	7 is a factor of 14
8 is a factor of 88	8 is a factor of 64
9 is a factor of 54	9 is a factor of 63

Challenge Master 18

In a set of three consecutive triangular numbers, if you find the product of the two 'outside' numbers, and square the middle number, the two results always have a difference which is the same as the middle number.

The difference between pairs of the squares of consecutive triangular numbers is always a cubic number.

Challenge Master 19

1. 952	**2.** 714	**3.** 1305
4. 684	**5.** 2040	**6.** 2628
7. 2272	**8.** 5967	**9.** 13 572
10. 15 036	**11.** 27 105	**12.** 35 136

Challenge Master 20

1. 3·7	**2.** 6·3	**3.** 1·3
4. 2·4	**5.** 2·1	**6.** 3·8
7. 0·6	**8.** 16	**9.** 17
10. 4	**11.** 23	**12.** 26
13. 66	**14.** 1·2	**15.** 36
16. 9	**17.** 2·4	**18.** 12

Challenge Master 21

1. £2·56 **2.** £3·68 **3.** £4·64 **4.** 5·32 **5.** £6·58
6. £3·08 **7.** £5·04 **8.** £3·12 **9.** £3·96

1. £2·72 **2.** £3·91 **3.** £4·93 **4.** £6·46 **5.** £7·99
6. £3·74 **7.** £7·14 **8.** £4·42 **9.** £5·61

Challenge Master 22

The number of ways relates to the factors of the number of squares. For example,

a 2 x 3 grid = 6 squares

factors of 6 are 1, 2, 3, and 6

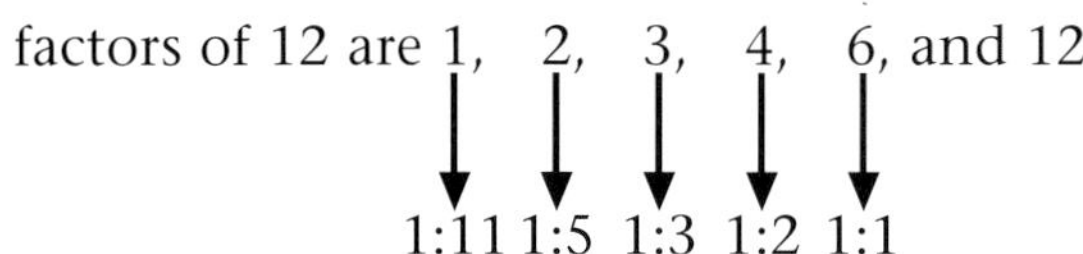

a 3 × 4 grid = 12 squares

factors of 12 are 1, 2, 3, 4, 6, and 12

1:11 1:5 1:3 1:2 1:1

Each factor, except the number itself, provides a way of shading so that the ratio of shaded to blank squares is 1 : X.

Challenge Master 23

1	2	3	4	5	6	7	8	9
2	4	6	8	1	3	5	7	9
3	6	9	3	6	9	3	6	9
4	8	3	7	2	6	1	5	9
5	1	6	2	7	3	8	4	9
6	3	9	6	3	9	6	3	9
7	5	3	1	8	6	4	2	9
8	7	6	5	4	3	1	1	9
9	9	9	9	9	9	9	9	9
10	2	3	4	5	6	7	8	9

Challenge Master 24

The happy numbers, up to 100 are:

1 7 10 13 19 23 28 31 32 44 49

68 70 79 82 86 91 94 97 100

Challenge Master 25

	Miles	Kilometres
Aberdeen	502	803
Birmingham	106	167
Bristol	115	184
Cardiff	159	254
Carlisle	301	482
Dover	71	114
Edinburgh	378	605
Exeter	172	275
Glasgow	397	635
Holyhead	259	414
Inverness	536	858
Leeds	189	302
Liverpool	202	323
Manchester	185	296
Newcastle	274	438
Norwich	114	182
Nottingham	122	195
Plymouth	210	336
Sheffield	160	256

Challenge Master 26

Possible units of measurement:

A acre
B bushel
C Centigrade, centimetre, centilitre, cubic centimetre, cubit
D day, decimetre, dram
E
F Fahrenheit, fathom, foot, furlong
G gallon, gram
H hand, hectare, hour, hundredweight
I inch
J
K kilogram, kilometre
L light year, litre
M metre, mile, millilitre, millimetre, minute
N nautical mile
O ounce
P peck, pint, pound
Q quart
R rod
S second, square centimetre, stone
T ton, tonne
U
V
W week
X
Y yard
Z

Challenge Master 27

shape	A	I	P
1	12	6	14
2	4·5	1	9
3	6	3	8

$A + 1 = I + {}^{P}/_{2}$

Challenge Master 28

o'clock times

1 o'clock, 11 o'clock 30°
2 o'clock, 10 o'clock 60°
3 o'clock, 9 o'clock 90°
4 o'clock, 8 o'clock 120°
5 o'clock, 7 o'clock 150°
6 o'clock 180°

half past times

12:30, 11:30 165°
1:30, 10:30 135°
2:30, 9:30 105°
3:30, 8:30 75°
4:30, 7:30 45°
5:30, 6:30 15°

quarter past times

12:15 $82\frac{1}{2}°$
1:15 $52\frac{1}{2}°$
2:15 $22\frac{1}{2}°$
3:15 $7\frac{1}{2}°$
4:15 $37\frac{1}{2}°$
5:15 $67\frac{1}{2}°$
6:15 $97\frac{1}{2}°$
7:15 $127\frac{1}{2}°$
8:15 $157\frac{1}{2}°$
9:15 $172\frac{1}{2}°$
10:15 $142\frac{1}{2}°$
11:15 $112\frac{1}{2}°$

quarter to times

12:45 $112\frac{1}{2}°$
1:45 $142\frac{1}{2}°$
2:45 $172\frac{1}{2}°$
3:45 $157\frac{1}{2}°$
4:45 $127\frac{1}{2}°$
5:45 $97\frac{1}{2}°$
6:45 $67\frac{1}{2}°$
7:45 $37\frac{1}{2}°$
8:45 $7\frac{1}{2}°$
9:45 $22\frac{1}{2}°$
10:45 $52\frac{1}{2}°$
11:45 $82\frac{1}{2}°$

Challenge Master 29

a. Reversing the vertical coordinates leads to reflection in the horizontal axis.

b. Reversing the horizontal coordinates leads to reflection in the vertical axis.

c. Reversing both coordinates leads to reflection in the horizontal axis, followed by reflection in the vertical axis. These are equivalent to a half-turn rotation about the origin (0, 0).

Challenge Master 30

Answers will vary.

Challenge Master 31

$3 \times 3 \times 3$
Number of cubes with:

0 red faces: 1

1 red face: 6

2 red faces: 12

3 red faces: 8

$4 \times 4 \times 4$
Number of cubes with:

0 red faces: 8

1 red face: 24

2 red faces: 24

3 red faces: 8

$n \times n \times n$
Number of cubes with:
0 red faces: $(n - 2)^3$
1 red face: $6\,(n - 2)^2$
2 red faces: $12\,(n - 2)$
3 red faces: 8

Challenge Master 32

1. square **2.** cube **3.** rectangle
4. hexagon **5.** cylinder **6.** rhombus
7. sphere **8.** parallelogram **9.** triangle
10. cone **11.** cuboid **12.** pentagon
13. octagon **14.** prism **15.** circle
16. pyramid

Challenge Master 33

The area of a kite is equivalent to half the product of the length of the diagonals.

Challenge Master 34

Answers will vary.

Challenge Master 35

Seaville: Mean house price £107 100
Oldtown: Mean house price £86 250

Challenge Master 36

d	**1**	**2**	**3**	**4**	**5**	**6**
1	0	1	2	3	4	5
2	1	0	1	2	3	4
3	2	1	0	1	2	3
4	3	2	1	0	1	2
5	4	3	2	1	0	1
6	5	4	3	2	1	0

d	0	1	2	3	4	5
probability	$\frac{6}{36}$	$\frac{10}{36}$	$\frac{8}{36}$	$\frac{6}{36}$	$\frac{4}{36}$	$\frac{2}{36}$